A SHEARWATER BOOK

Which World?
Scenarios for the 21st Century

*To Alice, my companion in the past and the present,
and to our son, Ross, and our daughter, Lily,
and still other generations that will come after,
to whom the future belongs.*

Which World?

Scenarios for the 21st Century

GLOBAL DESTINIES, REGIONAL CHOICES

ALLEN HAMMOND

ISLAND PRESS / Shearwater Books

Washington, D.C. / Covelo, California

A *Shearwater Book*
published by Island Press

Copyright © 1998 by World Resources Institute

First paperback edition published in 2000.

Library of Congress Cataloging-in-Publication Data
Hammond, Allen L.
    Which world? : scenarios for the 21st Century / by Allen L.
Hammond.
        p.   cm.
    "Global destinies, regional choices."
    Includes bibliographical references and index.
    ISBN 1-55963-575-4 (cloth) — ISBN 1-55963-576-2 (pbk.)
    1. Economic forecasting. 2. World politics—1989—Forecasting.
3. Twenty-first century—Forecasts.   I. Title.
HC59.15.H33 1998   98–13589
330.9'051—dc21   CIP

Printed on recycled, acid-free paper

Manufactured in the United States of America
10 9 8 7 6 5 4 3

Contents

Preface

The twentieth century has seen both the Russian Revolution and, seventy-five years later, the emergence of the genetic fingerprinting technique that proved fraudulent the claim of the woman once thought to be Anastasia, youngest daughter of the last Russian tsar. Outside my window I can see my small sailboat, made of high-tech plastic and equipped with Mylar sails—materials that literally did not exist even fifty years ago. I write these words with a laptop computer that weighs but a few pounds and yet is more powerful than the huge, room-sized machines common only twenty-five years ago. Truly, the world is changing very rapidly, and not just technologically. The Soviet Union, born in the Russian Revolution, has disappeared, and the political ideology of Communism—which once held half the world in its sway—is rapidly following the Soviet Union into oblivion. The world's population is nearing 6 billion people, more than three times the number who lived on Earth at the beginning of the century. Women, who at the beginning of World War I could not even vote in the United States, are becoming a decisive electoral force.

New technologies, the rise and fall (and sometimes collapse) of nations, swelling populations, an increasingly global economy, striking increases in literacy, and other profound social changes—all these are transforming our world with extraordinary speed. Tomorrow's world, in consequence, will be quite different from today's. Moreover, many of the choices that now confront human society have long-term consequences, such as potentially altering Earth's climate for centuries or extinguishing treasured life-forms—whether magnificent wild animals or valuable plants—forever.

Under such circumstances, making wise choices about the future is not easy. Indeed, most of us individually, preoccupied with the problems of today, don't think much about tomorrow. And collectively, society has little in the way of organized attempts at foresight—the short-term horizons of the quarterly profit statement and the next election all too often dominate economic and political attention. Yet even unconsciously, we are making choices, shaping the future. It is as if we were driving into the future at high speed, over uncertain ground, and without headlights: Might we collide with an unexpected obstacle or even drive right over an unseen cliff? Could better headlights—new insights into the future—lead society to change course so as to avoid emerging problems and secure a better future for generations to come?

This book is one result of a five-year research effort to explore these questions. The study, known as the 2050 Project, was organized by three major research organizations—the Brookings Institution, the Santa Fe Institute, and my own institution, the World Resources Institute—and involved dozens of scholars from all over the world. As its name implies, the study focused on the next half century, the period between now and the year 2050.

The 2050 Project viewed human societies and their interactions with one another and with Earth as what scientists call a complex system—a system that is subject to abrupt shifts from one pattern of behavior to another. Studies of such systems show that traditional scientific approaches—simplifying the system or attempting to analyze only one aspect of it—can give very misleading answers: strong linkages among different parts of the system are what create its complexity and determine its often unpredictable behavior. To gain insight into a complex system, particularly one that is imperfectly understood, it is often necessary to take "a crude look at the whole," to use Nobel laureate Murray Gell-Mann's apt phrase.[1] Taking a crude look at the whole—considering demographic, economic, technological, environmental, social, cultural, political, and other factors that may determine the future—became a guiding principle for the 2050 Project, and for this book.

The 2050 Project set out to find paths or trajectories into the future that might lead society toward a favorable destination half a century from now, and it developed scenarios to explore future trajectories. This book draws on those scenarios and the underlying analyses; it also adapts scenarios developed by the Global Scenario Group—an independent international group—and by other independent scholars.

In addition to the 2050 Project, this book builds on my own decade-long experience in studying global trends. As it happens, for much of that period I served as editor-in-chief of the World Resources Institute report that summarizes the United Nations's environmental data for scholars and policy makers around the world.[2] Seeing these numbers cross my desk gave me a unique opportunity to analyze environmental, economic, and social trends in nearly 200 countries and to consider what the data reveal about potential conditions in the year 2025 or 2050.[3] Building on these data and the analyses of them done by me and my colleagues at the World Resources Institute and by many other scholars, this book asks where humanity appears to be headed and what destinations we could plausibly reach within the next half century.

Finally, this book incorporates information on many different regions and insights into their unique characteristics. Some of these insights come from my personal experience in different regions, but far more come from the experiences and knowledge of colleagues who are native to the regions or who have lived or worked extensively there. In addition, I sought out regional experts and scholars; I consulted the field reports of anthropologists and other professional observers stationed in one country or another under the auspices of the Institute of Current World Affairs; I used informal networks of journalists to find and interview people who exemplify the changes under way in particular regions; and I drew on regional workshops organized by the 2050 Project or one of its sponsoring institutions.

The result has been an intellectual odyssey that has greatly enriched my own appreciation of the world we live in, sharpened my concern for its future, and stimulated new ideas about how we might shape the future for the better. *Which World?* shares the insights of that journey with a wider audience, portraying three different but plausible futures for human society and exploring their implications.

One important caveat should be mentioned here. The world is inherently unpredictable, and truly unexpected events could occur. For example, scientists cannot rule out a sudden, disastrous shift in climate that would plunge part of the world back into a mini–ice age—a shift that actually happened once before, about 11,000 years ago. A terrible new disease—perhaps even one deliberately constructed as a weapon of biological warfare—could devastate society, as did the bubonic plague in the Middle Ages. An asteroid could strike Earth in a cosmic collision like the one that killed off the dinosaurs. Such things are possible, but in this book I focus on much more plausible events, much less extreme futures. Even so, as I describe on the pages that follow, human destiny remains deeply uncertain: for the next half century, there are both ample causes for concern and ample reasons for hope.

Allen Hammond
Rolphs Wharf, Maryland
30 November 1997

Acknowledgments

This book builds on and borrows extensively from the work and thinking of many of my colleagues who participated in the 2050 Project, my colleagues in the Global Scenario Group, and my colleagues at the World Resources Institute, as well as the publications of a large number of other scholars and analysts. I benefited from the help of journalists, scholars, and friends in every region of the world, many of whom contributed examples and expert local knowledge. But the synthesis and the interpretation are my own, and so must be the responsibility for what is printed on these pages.

In particular, I would like to acknowledge John Steinbrunner of the Brookings Institution, Bruce Murray of the California Institute of Technology, and Rob Coppock of the German–American Academic Council as comrades-in-arms during the development of this book, from whom I learned in ways large and small. Paul Raskin and Gilberto Gallopin of the Stockholm Environment Institute contributed enormously to my thinking on scenarios. Many other colleagues read and commented on the manuscript at various stages or helped me think through pieces of it, including Jan Clarkson, Michael Cohen, Ann Florini, Tom Fox, Sumit Ganguly, Jonathan Lash, R. K. Pachauri, Walt Reid, Bob Repetto, Veerle Vandeweerd, and Changhua Wu.

To Gus Speth and Jessica Matthews I owe my involvement in the analysis of global trends and many insights into the development process, and to Murray Gell-Mann I owe the initial inspiration for the 2050 Project that led to this book. Many other colleagues at the World Resources Institute inspired and helped in more ways than I could list. Sharon Bellucci, Philip Howard, and Carolina Katz helped in different phases of the research for the book; Maggie Powell helped with the figures.

Susan Sechler, then at the Pew Foundation, contributed both by pushing me to make the book better and by arranging funding. Chuck Savitt of Island Press believed in this book for years before it was written, and my editor there, Laurie Burnham, helped enormously to shape its structure and sharpen its prose. Finally, a thanks to those institutions whose resources made it all possible: the John D. and Catherine T. MacArthur Foundation and the Howard Gilman Foundation for supporting the 2050 Project and the Pew Charitable Trusts for supporting the writing of this book.

PART I

LOOKING AHEAD

Chapter 1

Thinking About
the Future

IN THE SPRING OF 1995, I taught a graduate class on the potential implications of long-range trends in population, economic growth, and the environment. It was a new experience for me—I'm a researcher and a writer, not a teacher—but what was really unexpected was the ingrained pessimism of my students, most of them knowledgeable professionals in their thirties. They seemed predisposed to believe that growing populations in poor regions of the world would inevitably lead to disaster, that environmental degradation could only get worse, that violence and conflict would inevitably escalate—in short, that a familiar litany of dark prophecies would come true. And they found it curious, even incomprehensible, that I, who knew the trends better than they, did not share their view.

It's not hard to see why my students and many others hold pessimistic views. An unprecedented 90 million people are added to the planet each year, most of them in the poorest countries, which are least able to accommodate them. Poverty, disease, and hunger continue to blight the lives of hundreds of millions of people. Valid concerns persist about sweeping global economic changes that could eliminate jobs and livelihoods, undermining whole communities; about rising economic disparity; about failing governments and worsening social conditions. It's certainly possible that this generation's legacy to the next will be an Earth poisoned by industrial toxins, shorn of virgin forests, and committed to an altered climate. Every week, it seems, there is fresh evidence of a world in trouble—another African country in chaos; a killer smog that shuts schools and airports in several Southeast Asian countries; new violence or drug-related corruption in Latin America. Yet as I hope to show, the prospects for the future are more complex, and ultimately more hopeful, than such headlines suggest.

Just as I am troubled by simplistic pessimism, however, I find the other extreme, simplistic optimism, even more disturbing. True, there is much to be optimistic about: the spread of democracy and market economies, the rapid advance of new technologies, widespread improvements in literacy. The peaceful evolution of South Africa into a multiracial democracy, the remarkable and unheralded introduction of village-level democracy in China, and the rapid spread of economic reform in Latin America suggest that positive changes are under way. It is certainly possible that such trends will dramatically increase opportunity, wealth, and human welfare, at least for many of the world's people, and that new knowledge and human ingenuity will engender solutions to many social and environmental problems. The next half century might really see the emergence of the world's first truly global human civilization. But the operative words here are "might" and "possible."

Might Latin America, for example, overcome its tradition of neglect for the poor and special privilege for the very rich that makes it the most inequitable region on Earth? That will require far more than economic reform.

Will it be possible for China—the world's fastest-growing market economy, embedded in the socialist structures of the last major Communist power—to defy its internal contradictions and tensions, its massive pollution, and its surging urban migration? Or will its problems overwhelm it and dissipate its momentum, as has occurred, at least temporarily, in Southeast Asia?

Might the other forty-odd nations of sub-Saharan Africa follow South Africa's lead? Or will stability and competent governments come too late to avert a downward spiral of environmental degradation, malnutrition, impoverishment, and possibly widespread violence and chaos?

And will the world's wealthy nations accept the challenge of open economies and global leadership? Or is it possible that, preoccupied with domestic social issues and aging populations, they will turn inward? Of all regions, North America, Europe, and Japan have the fewest internal constraints on their futures—but the most to lose if a larger world turns desperate, unstable, and polluted.

Does it matter if some countries or some regions succeed while others fail? I believe it does. The world is already so strongly interlinked that no country stands alone; no region's future can be fully separated from that of others.

Examples of such linkages are multiplying. In the fall of 1997, a currency crisis in Southeast Asia triggered a crash in stock markets around the world, illustrating that with tightly coupled markets, imprudent financial choices locally can have global consequences.

Health concerns are becoming global in scope. The HIV virus that causes AIDS threatens every region, and the deadly Ebola virus is only twenty-four hours and an international plane ride away from anywhere on Earth. Thus, growing human activities in African forests—the reservoir from which HIV sprang and which harbors Ebola—can have a worldwide impact.

Energy provides still other linkages. In coming decades, the world will depend more and more on the oil reserves of the Middle East, and tensions there—between Israel and its Arab neighbors, between autocratic rulers and the rising forces of radical Islam—seem on the rise, putting global energy supplies at risk. And China, which plans to fuel its rapid industrialization with its huge reserves of coal, will soon become the largest source of the greenhouse gases that can cause global warming. Such choices, and those of the industrial regions that are already huge greenhouse gas emitters, may help shape the whole planet's climatic future.

The world is also connected by human movement, with illegal migrants from impoverished and chaotic regions overflowing borders everywhere, reaching levels that provoke strong social and political reactions. Thus, regional choices that bring economic growth or stagnation, that eliminate poverty or let it persist, that foster stability or chaos and violence, can have repercussions far away.

Even crime now has global dimensions. Police agencies report that the violent Russian mafia is linking up with Colombian drug cartels to create a powerful global criminal alliance capable of moving arms, drugs, and money across borders in massive quantities, spreading violence and corruption in rich and poor regions alike, and subverting private firms or perhaps whole governments.

In such a tightly coupled world, we are all neighbors. Failures create wide ripples, and misery tends to travel. No region, consequently, can be entirely the master of its own fate: the global destiny depends on regional choices made separately in many different corners of the world.

At the dawn of the twenty-first century, the challenge for the human race is no longer primarily surviving the onslaught of natural forces, winning a living from an often harsh and unpredictable environment, as it has been for most of human history.[1] Nor is it the titanic struggle between the two opposing economic and political systems that overwhelmingly shaped the last half of the twentieth century. Today, humanity faces a fundamentally different challenge—that of managing a planet and a global human civilization in ways that will sustain both indefinitely.

What makes this task less than easy are the pace and complexity of change. Over the next half century, human society will undergo a profound demographic transformation, experience fundamental shifts in the global balance of economic and political power, and cope with nearly continuous technological change. These transformations are inevitable—the forces that compel them are already in place—but their outcomes are far from fixed.

No one can say for certain whether the world is headed for better times or worse. Nor can we know whether the twenty-first century will bring new heights for human society or conflict, degradation, and human tragedy on a scale that overshadows even the excesses of the twentieth century.

Is it within our power to tip the balance toward a future world that we would *want* to leave to our descendants? Some would argue that our destinies are in the grip of larger forces, that there is little we can do to shape the future. Others are simply indifferent, sure that the world will muddle through somehow. But still others—and I share this view— think that neither passivity nor complacency is good enough because human actions have the capability to shape tomorrow's world as never before in history, for better or for worse. To decide which actions are critical, however, requires that we know more about what the future may hold.

This book is about the future, but not in the sense of making predictions. Rather, it suggests how to think about the future. Because human destiny is not predetermined, this book explores not just one but several possible worlds, each embodying a very different vision of the future. Implicit in these contrasting visions is a choice: which world do we prefer; which world do we want to pass on to our children and grandchildren?

To inform such choices, to engender the insights on which to base our actions, this book offers a three-pronged analysis. First, this book examines long-term trends. Despite the mind-numbing complexity of our era and the glut of information that deluges us daily, it is possible to sort out the factors that will matter most in shaping the future— those, for example, that drive the transformations mentioned earlier. These critical trends, traced forward globally and region by region, illuminate the possibilities that lie ahead. Second, although we cannot know the future, we can envision it. So this book describes and compares different trajectories or scenarios that society might plausibly follow—scenarios that lead to radically different worlds and that may shed light on the social choices that might distinguish one path into the future from another. Third, and perhaps most important, this book looks at the world region by region, combining critical trends, scenarios, and information on cultural, social, and political context. Because regional choices define our interlinked global destiny, the possibilities and unique constraints of each region are crucial to comprehend.

Long-term trends form the bedrock of thinking about the future. Some of these trends are positive. For example, the rising efficiency of industrial processes and the changing structure of industrial economies are reducing pollution. Producing software is now a large part of the U.S. economy and employs far more people than does making steel, drawing on knowledge rather than mineral ores as the primary resource. An important social trend in virtually all developing countries is declining birthrates and dramatically rising contraceptive use—from less than 10 percent of married couples in the 1960s to more than 50 percent in the 1990s. Both of these trends seem likely to continue in coming decades.

Other trends are not so encouraging. Nearly a third of the people in sub-Saharan Africa don't get enough to eat, and the number is growing. Urban populations are exploding in developing countries, with close to a million migrants per week pouring in from the countryside—far faster than decent housing can be built or water systems expanded. Violence also appears to be on the increase: kidnapping and drug violence here, religious conflict there, domestic terrorism even in Japan and the United States, armed banditry in many places, organized and increasingly violent criminality worldwide. To an alarming degree, the signs suggest that the world is moving toward a troubled future.

But trends cannot be the only guide to the future, because the unexpected can occur, producing both happy and unhappy surprises. Moreover, there are many important phenomena so volatile that no long-term trend can be charted, let alone projected into the future. Social attitudes can shift dramatically, as demonstrated by the startling reversal in U.S. attitudes toward smoking over the past ten years. Or consider how rapidly stock market investors can switch from optimistic to pessimistic views of the future, sending markets plunging, or how quickly seemingly invincible political mandates and electoral advantages fall apart when voters change their minds. As investors and political analysts know, changes in such phenomena are inherently difficult to forecast—few if any experts foresaw the collapse of the Soviet

Union—yet they can profoundly alter a society's social or political context. Thus, even in a country where the trends in birthrates, economic output, and environmental degradation look dismal, a new political consensus can arise suddenly, bringing with it a radical change in that country's prospects. An example is the abrupt transformation in Poland and the Czech Republic in recent years.

If, indeed, changes in social attitudes or shifts in behavior are fundamental drivers of the future, how can we take them into account or explore the consequences of a new political consensus and the changes in policy that might flow from it? Herein lies the power of scenarios—precisely constructed stories about the future that describe alternative futures or contrasting trajectories.

Scenarios serve as stimulants for our imagination. They help us to conceive of new possibilities, to explore wildly different alternatives, and to integrate many different factors into our thinking about the future. Thus, we can *imagine* a future constrained by environmental degradation and resource limits—in which widespread poverty, desperate shortages of food, and huge income gaps between rich and poor cheapen the worth and dignity of human life—or one set free of such constraints by new technologies and wiser policies, a future in which all people's basic needs are met and the bounty of a global industrial civilization is widely shared.

Scenarios also offer a means to explore some of the critical choices that will, or could, influence the future. Constructing a number of different scenarios often highlights actions and strategies that might enlarge society's options or increase the likelihood of more desirable futures.

Critical trends and scenarios constitute vital parts of the tool kit we will use for thinking about the future, but they are not sufficient in themselves. In my travels to other parts of the world, I am often intrigued by how different one country is from another. Yet with all the media coverage of global trends and global markets, it is all too easy to think of the world as a single place, forgetting that conditions—and, more important, aspirations and cultural patterns—vary enormously. Indeed, I believe that a fundamental flaw in many studies of the future is that they look only at global patterns or global scenarios—in effect, they treat the world as a homogeneous unit. Regional differences powerfully constrain not just what the critical problems are but also how the problems are perceived and how solutions must be sought. Thus, regional differences must play a central role in explorations of our common destiny.

From a distance, for example, Africa and its problems may seem hopeless, but few people realize that the continent is still relatively uncrowded, with a lower population density than the United States. Africa has more minerals, more fertile land, and more water per person than either China or India—and this will be the case even when its present population has doubled and is roughly the same size as China's. When the region shakes off the "hangover" of its colonial period, which ended scarcely a generation ago, it may emerge as a continent of promise, a full participant in the global market. Is the hopeful experiment in multiracial democratic government now under way in South Africa the harbinger of that transformation?

And what of China? Today, media reports make China's future as an economic superpower and a global political force seem assured. But more than once in its long history, China has been convulsed by civil war and fragmented into several nations; could such a transformation happen again as an aging leadership and an ever weaker central government confront divergent regional interests? Can China maintain social stability in the face of what appear to be unprecedented levels of rural-to-urban migration, with perhaps 250 million new residents expected to move into its cities in the next fifteen years? And if China falters, what effect will its instability have on the rest of the world?

Looking to Latin America, the future seems uncertain. On one hand, the region seems poised for economic growth: most governments are democratic; there is a widespread consensus for economic reform; its natural resources are richer relative to population than are those of any other region; its industrial output is larger than that of China. On the other hand, Latin America has the most concentrated ownership of land and the most disparate incomes of any region in the world. Homeless children flood the streets of São Paulo, Brazil, and other cities. The endemic corruption of the cocaine trade and the spread of violence—from guerrilla bands, from unchecked criminal groups, from those embroiled in rural conflicts over land and urban conflicts among slum gangs, from unofficial death squads—also weigh heavily on the region. Could such social tensions, combined with the failure to share the benefits of growth more broadly, undermine the region's prospects and again lead to instability and economic stagnation?

India, too, seems poised for economic growth. With a vibrant democracy and a well-educated middle class, it has a head start over many developing regions. But most of its people live in poverty even more desperate than Africa's—whether measured by income, illiteracy, or lack of access to basic needs—and its population is likely to surpass that of China. Conflicts based on religion and caste are rising. Limited economic reform has begun; will social reform follow, and will it come in time?

Prospects for North Africa and the Middle East are equally uncertain. This oil-rich but arid region faces a water crisis: by 2025, thanks to rapidly rising populations, the demand for water in the region will be four times what nature can provide. But the region faces a more critical challenge, namely, social and political reform. Left unchecked, social pressures will continue to build, making revolutionary upheavals more likely, with fundamentalist Islamic groups virtually the only alternative to authoritarian governments. Yet such groups, if they come to power, are unlikely to advance the social and political changes—including changes in the status of women—that will be required to stabilize populations and to modernize economies. The likely result would be continued instability and conflict and potentially major impacts on a world increasingly dependent on the region's oil.

And what of North America, Europe, and Japan? Despite their present position as the world's dominant economic powers, all three regions face oddly uncertain futures. At stake is whether these regions will use their enormous economic and social advantages and their command of capital and technology to lead the world toward more hopeful trajectories or whether they will give in to reactionary social and political tendencies, already evident in all three regions, and forge a narrower, inward-focused strategy.

Among all the contrasting possibilities, which trends will prevail, which scenarios unfold? I argue that if current trends persist, the future will combine both pessimistic and optimistic elements: Wealth and opportunities will increase for some, perhaps even for many, of the world's people, but the world is likely to become a far less desirable place to live in for most, perhaps even for the rich. New solutions to problems will appear, but they may not be put in place in time to prevent widespread environmental degradation. The twenty-first century may indeed see an unprecedented global civilization bound together by the Internet and its successors, but it may also see human tragedy on a scale that could make the Holocaust seem modest.

But I also argue that as human beings, we can choose our destinies, to a large extent, and can steer our collective enterprise toward any one of several worlds. Thus, I am fundamentally optimistic: I argue that no future is inevitable and that many of the negative trends could be reversed if society chooses to do so and can summon the will to act. Moreover, I believe that fundamental social and political changes that could shape a better world are already under way, if only tentatively so, although these beginnings are largely unappreciated.

Most important, I believe that if we are to shape desirable futures for our children and grandchildren, we must know more about where the world appears to be headed and what choices we need to make, collectively and individually. This book is intended to help in that process—to serve as a primer on the future, a guide to what might be . . . and what could be.

Chapter 2

The Power of Scenarios

MAKING CHOICES ABOUT THE FUTURE means coping with bewildering complexity and uncertainty. In ancient times, our ancestors often sought guidance in stories, such as those that give rise to the myths and legends found in every culture. A more modern method for grappling with uncertainty is to construct scenarios—carefully posed stories that describe plausible alternative futures, often supported by reams of data and the intuitions and experience of experts and scholars. In recent years, business executives and military planners have often turned to scenarios precisely because they offer a powerful tool for thinking about choices.

Scenarios are not predictions or forecasts. Rather, they suggest how the world *might* turn out. Like any good story, scenarios have a set of circumstances or constraints within which they occur; a plot or logic that guides how events unfold, and characters—individuals or groups or institutions—that take part in the events and give them a human context. But the point of the story is not to develop a more accurate view of the future; it is to enable the reader to look at the present in a new light—to see it more clearly, to see new possibilities and implications—and hence to make better decisions.

Scenarios are powerful because they help those who read them to visualize in more concrete and human terms—to emotionally connect with—what might otherwise be only an abstract trend, a line on a graph. They make far more vivid the potential consequences of current trends or proposed actions. They also can challenge assumptions that are so deeply held we may not be aware of them and thus help free us from the grip of the past and the present. Most of us, after all, march into the future looking backward, guided by the accumulated lessons of past experience; scenarios enable us to look forward, to prepare for a future that will assuredly be different from today.

The commercial world offers many examples of scenarios that have affected billion-dollar decisions. Peter Schwartz, one of the most accomplished scenario builders of recent times, tells in his book *The Art of the Long View* how scenarios helped Royal Dutch Shell Group become one of the most successful of the major oil companies.[1] In the early 1970s, for example, Shell's Group Planning Department constructed scenarios to help the company's senior managers predict the effects of a drastic increase in the price of oil. The planners found that it was not enough simply to suggest future possibilities; managers, firmly in the grip of decades of past experience with low oil prices, simply didn't respond. Instead, the Shell scenario team found it necessary to change the managers' view of reality, their mind-set, by vividly describing "the full ramifications of possible oil price shocks" and hence enabling them to *feel* what it would be like to live and make decisions in such a future.[2] When the 1973 oil crisis occurred and prices did rise dramatically, Schwartz says, of the major oil companies only Shell "was prepared emotionally for the change." The company prospered and became the second largest of the global oil giants. Scenarios also enabled Shell to anticipate the 1986 collapse of oil prices, gaining the company additional commercial advantages.

Scenarios are widely used by military planners and the intelligence community. In 1996, the U.S. Department of Defense undertook a scenario-based study of the security risks posed by China's emergence as a major economic power and by other plausible changes in eastern Asia. One scenario asked what might happen if China demanded unification with Taiwan, using long-range missiles to destroy Taiwan's power plants and blockading its ports; another considered Chinese expansion into oil-rich regions of Siberia and central Asia; a third contemplated the unification of North and South Korea. Such scenarios force military strategists and security experts to consider extreme events that might normally be overlooked and to focus on critical factors such as nationalistic sentiment that might distinguish one future from another.

Scenarios can also help commanders plan tactical military operations and choose equipment for their troops. Before the North Atlantic Treaty Organization (NATO) forces entered Bosnia, British defense specialists used explicit (but still classified) scenarios to help plan operations for their troops—scenarios that British sources say significantly changed the way operations are managed on the ground in that war-torn country.

The U.S. intelligence community and the U.S. Department of State recently collaborated on scenarios to aid in choosing antiterrorist policies. They created four distinct futures: an integrated global economy; a world divided into competing trade blocs; a fundamental split between the industrial and the developing worlds; and a renewal of East–West cold war tensions. Then they assembled a group of intelligence analysts, military experts, diplomats, and academic specialists, divided them into four teams—one to each scenario—and asked each team to analyze the terrorist threats consistent with its scenario. As a result, planners in the State Department and in U.S. intelligence agencies are sharpening their thinking about terrorist threats and planning counterterrorism activities.

The broader business community is also turning to scenarios. More than thirty major corporations recently undertook a joint scenario study under the auspices of the Geneva-based World Business Council for Sustainable Development.[3] The goal was to develop a common set of scenarios that could help corporate leaders anticipate and respond to a world of increasing environmental and social demands, thus putting them ahead of the curve.

Although scenarios have been employed primarily by military planners and corporate strategists, there is no reason why they cannot be turned toward wider social purposes and incorporated in the public discourse. Not to do so, in fact, is to leave these powerful tools in the service of narrow interests. Global corporations and intelligence agencies, after all, are not the only ones that have a stake in the future. We all do, either directly or through our children and grandchildren. "Public policy," says Ged Davis, an executive with Shell International Ltd., "must be not only adaptive but also anticipatory."[4] Scenarios can help.

To be useful, however, a scenario must jar us out of familiar assumptions and challenge us to think about how the world might be different. It must therefore be powerfully, even provocatively, stated. One technique is to write a scenario as a "future history," as if a timetraveler had brought back an account written years in the future. Consider a portion of a scenario that explores what might happen if current environmental trends were to continue for another half century, written from the perspective of the year 2050.

Oil is no longer much of a geopolitical factor today; even the vast Saudi reserves could not keep up with Asia's huge energy needs, and earlier in the century prices surged as chronic shortages developed, eventually forcing a global shift to natural gas and, increasingly, to renewable forms of energy such as solar power, wind power, and biofuels. Renewable sources of energy now account for a third of global energy supplies.

Despite that shift, environmental conditions are worse than they were half a century ago. The climate is distinctly warmer and more variable and is expected to become more so. Floods are so common that many of the low-lying areas along the Mississippi River and other major U.S. rivers have been abandoned, with farms and whole towns lost, and repeated droughts plague Australia and parts of Africa. In some of the small island countries now at risk of submersion by rising seas, preparations for evacuation are under way.

Climate isn't the only concern. Despite urgent attempts now being made to preserve threatened ecosystems, for many it is too late. Living coral reefs survive in only a few remote atolls, and there are no large tracts of tropical forest left, other than commercial timber plantations and the (much reduced) natural stands in South America's Amazon basin. Crop failures are common, reflecting a shortage of wild cultivars from which to breed plants resistant to new diseases and capable of coping with changed climatic conditions. Air and water pollution is an urgent, universal concern throughout newly industrialized Asia and Latin America.

People have adjusted, of course. Electric and other new-generation automobiles now predominate in most urban areas because pollution taxes and high oil prices have made gasoline and diesel vehicles prohibitively expensive. Air travel is much less common, its high ticket prices dictated by the soaring costs of aviation fuel; in fact, "virtual" tourism has become the vacation preference of many. But there is a sense of irretrievable loss.

A scenario scripted by an environmental group? Not exactly. This future is based on a long-range energy scenario developed by Royal Dutch Shell;[5] on the consensus report of an international panel of more than a thousand leading climate scientists;[6] and on recent data about the status of forests, coral reefs, and other critical ecosystems.[7] Of course, scientists don't know how severe the effects of climate change and other forms of environmental degradation will be fifty years from now or just how societies will adapt, but many scientists would find such a scenario plausible if current trends continue.

Trends can change, of course. If the world comes to believe that severe environmental degradation is likely, steps can be taken to prevent such a calamity. Even when prevention is not possible, scenarios can play a powerful role in helping society anticipate and hence be more prepared for possible changes, just as Shell foresaw first the rise and then the collapse of energy prices. But environmental change is not the only challenge the future may hold; another concern—one I address in more detail later—is the failure of social stability.

In retrospect, the early signs of Africa's disaster were evident in the ethnic warfare and sporadic government collapses that began

in the 1990s. Over the next two decades, with rapidly rising pop-
ulations, falling incomes, decaying roads, and corrupt and self-
serving governments, many Africans came close to the brink of
destruction. Food became increasingly scarce—more than half the
population was malnourished—and crime was all too often the
only way to feed a family. The collapse into near anarchy began in
West Africa as state after state foundered, but it spread across cen-
tral Africa and then into East Africa like a virulent plague. The
resulting flood tide of refugees overwhelmed the few stable
nations that remained and threatened southern Africa as well.
Zimbabwe and South Africa resorted to shooting illegal migrants
at the border. Their sheer numbers, along with widespread chaos
and violence, overwhelmed aid agencies and made it impossible to
bring in food and medicines; after a few halfhearted attempts,
Western nations abandoned efforts to intervene. At the height of
the troubles, more than 5 million people per year died of violence,
hunger, and disease. Only with the rise of a new generation of
leaders, who imposed order and began to reconstruct African soci-
ety, did the troubles cease.

Although the future pictured in this scenario fragment is only
conjecture, the trends it describes are real enough, incorporating data
from the World Bank, the Food and Agriculture Organization of the
United Nations, and other sources. Indeed, the possibility of widespread collapse in Africa—dramatically pictured by journalist Robert
Kaplan, whose *Atlantic Monthly* article[8] captured the attention of the
Clinton administration—is plausible enough to worry development
organizations concerned with the region as well as many thoughtful
Africans.

For all the pessimism, however, there is a more optimistic end of the
spectrum, a world that some visionaries foresee arising from today's
explosion in information technology. In such a "wired" world, digital
communication links connect every community on the planet (even if
many of the links are cellular or satellite borne). Like the preceding scenario fragment, this too is based on real trends and is addressed in more
detail later.

The GlobalNet now provides access to a billion times as much information, and in a far more organized fashion, than did the fledgling Internet of the year 2000. People do most of their business and financial transactions on-line, and there is near universal electronic access to government and business services. Shopping is a global affair, with crafts, delicacies, and other products easily viewed and ordered from anywhere. Moreover, the GlobalNet has become the nervous system both of local communities and of our global civilization, binding together far-flung families, friends, clubs, and citizen's organizations, linking worldwide social coalitions and corporate alliances, and in the process transforming governance and politics everywhere. Even remote villages are no longer isolated.

Perhaps the greatest impact of the GlobalNet has been on the developing regions of the world. It is hard to imagine what life was like in such regions before every village had access to educational programs and teaching software or to instant medical consultations; before farmers in even the poorest countries could get crop advice and long-range weather forecasts as well as local market prices; before village cooperatives and individuals could get loans, even for small amounts, via remote credit plans. By bringing such services to villages, the GlobalNet has revolutionized rural society, exposing billions of individuals to wider horizons and new opportunities. In a single generation, perspectives have shifted to a more modern social and political outlook, even in countries that still lack an industrialized economy, in the process accelerating economic and social development.

With international corporations such as Teledesic and Globalstar building new global satellite telecommunications networks, the technical infrastructure envisioned in this scenario may well exist within the next quarter century, perhaps even sooner. Whether interactive electronic networks will profoundly affect societies in the way described, and whether people and societies will make use of this new capability in such novel ways, is much more speculative. However, the experiences of U.S. communities that have universal access to electronic mail and those of Finland, where on-line access is further advanced than in any other country, certainly suggest that there is potential for accelerated social and political development.

As the preceding scenarios illustrate, the future holds an immense range of possibilities. Is it possible to construct a set of scenarios that encompass such diversity, that incorporate and fairly represent the conflicting world views, the philosophical divides, of our era? Might such scenarios provide a set of images and a shorthand language that would aid coherent discussion about the future? What might these imagined worlds look like?

PART II

THREE WORLD VIEWS

IN NOVEMBER 1995, a small meeting took place in the converted townhouse offices of the Tellus Institute, just off the historic Common in downtown Boston. Although there were fewer than a dozen participants in all, the group was remarkably diverse, with representatives from five different continents—among others, a lawyer from Africa, the eloquent head of a nongovernmental development organization in Bangladesh, a passionate Latin American scientist and a methodical Dutch one, as well as two North Americans. The group met at the behest of Paul Raskin, a U.S. physicist turned energy analyst and scenario builder, on behalf of the Stockholm Environment Institute. The diversity and breadth of experience were deliberate because the goal of this group was nothing less than to design a set of global scenarios that span many different ways of looking at the world. Thus was born the Global Scenario Group, of which I am privileged to be a member.

The discussions at the Boston meeting were intense, because the group explored not only the members' different ideas but also their often very different ways of looking at the world. Gradually, however, from a bewildering number of possible futures put forward, a framework emerged for thinking about the future. From that meeting and more than a year's subsequent work came a set of scenarios that reflect three altogether different visions or world views. These are, respectively, a *Market World,* in which current patterns continue; a *Fortress World* that reflects fundamental but undesirable social change; and a *Transformed World* that reflects fundamental and desirable social change.[1] The first of these worlds is hardly static; it assumes continued "evolution, expansion, and globalization," but without major surprises or changes in direction. The second two, however, envision "profound historical transitions over the next century in the fundamental organizing principles of society"—for the worse and for the better, respectively.[2] My adaptation and simplification of the Global Scenario Group's scenarios, summarized here and explored in more detail in the following chapters, form an essential building block for the rest of this book. My versions incorporate materials from other sources, too, but at their core they are built on the collective imagination and experience of the Global Scenario Group, a debt I am glad to acknowledge.

Market World

In this scenario, economic reform and technological innovation fuel rapid economic growth. Developing regions are integrated into the global economy, creating a powerful global market and bringing modern techniques and products to virtually all countries. The result is widespread prosperity, peace, and stability.

Market World is a compelling vision, endorsed implicitly or explicitly by many corporate leaders and economic theorists and seemingly justified by the failure of centrally planned economies. Its voice can be heard on the editorial pages of the Wall Street Journal and the Economist. What could deny this future or dim its promise? And what, in this post-Communist era, are the alternatives to this vision?

Fortress World

At least two broad alternatives to Market World emerge. One, Fortress World, is a fundamentally pessimistic vision based not only on the failure of market-led growth to redress social wrongs and prevent environmental disasters but also on the belief that unconstrained markets will exacerbate these problems and that large portions of humanity will be left out of the prosperity that markets bring. These failures eventually destroy the resources and the social framework on which markets and economic growth depend. Economic stagnation spreads as more resources are diverted to maintain security and stability, as does economic fragmentation where conflict dominates or the social order breaks down. The scenario describes the dark side of global capitalism, a future in which enclaves of wealth and prosperity coexist with widen-

ing misery and growing desperation, a future of violence, conflict, and
instability.

Transformed World

A second alternative is a visionary scenario in which fundamental social
and political change—and perhaps even changed values and cultural
norms—give rise to enlightened policies and voluntary actions that
direct or supplement market forces. *Transformed World* envisions a
society in which power is more widely shared and in which new social
coalitions work from the grass roots up to shape what institutions and
governments do. Although markets become effective tools for economic progress, they do not substitute for deliberate social choices; economic competition exists but does not outweigh the larger needs for
cooperation and solidarity among the world's peoples and for the ful-
fillment of basic human needs. In effect, this optimistic vision asserts
the possibility of fundamental change for the better—in politics, in
social institutions, in the environment.

These three scenarios are deliberately constructed to frame sharply contrasting images of the future. In reality, the world in 2050 is likely to
contain elements of all three scenarios, especially when the divergent
prospects of different regions are considered. But the scenarios nonetheless provide a convenient shorthand for widely held but contrasting
visions of human destiny and serve as a useful framework for analysis.

The scenarios are also intended to be clearly within the realm of the
plausible—they are based on existing trends. Of course, more extreme
futures cannot be entirely ruled out; completely unexpected events
could occur. An asteroid might collide with Earth, devastating life
here—it's possible, even if not likely—or an unforeseen climatic disaster could doom whole regions, plunging Europe back into an ice age,
for example. Or a scientific breakthrough could create sudden new possibilities—even though, as we will see, it usually takes many decades for
even profound discoveries to have widespread social effects. But even

leaving such extreme, almost science fiction–like, futures aside, the scenarios suggest that humankind still faces starkly different possibilities:

- Optimistic futures in which economic and human progress occur almost automatically or with modest economic reforms, driven by the liberating power of free markets and human initiative.

- Darker futures in which unattended social and environmental problems threaten progress and diminish its promise, bring rising conflict and violence, and doom hundreds of millions of human beings to lives that are, in Thomas Hobbes's famous phrase, "nasty, brutish, and short."

- Futures that transform the human condition—that offer a better life, not just a wealthier one, and that seek to extend those benefits to all of humanity.

The *Market World* vision is appealing to many people—it would be nice if the future would just take care of itself. But can largely unguided market forces realistically bring us closer to the kind of future we want? Could things really plummet to the degree pictured in the *Fortress World* scenario? Are the more fundamental changes implicit in *Transformed World* really needed? And what would these fundamentally opposed visions of the future mean for our lives and those of our descendants?

Chapter 3

Market World: A New Golden Age of Prosperity?

"For the foreseeable future, the only international civilization worthy of the name is the governing economic culture of the world market."
—*Richard Rosecrance*[1]

TO MANY CORPORATE CHIEFTAINS, high-tech entrepreneurs, and political conservatives, the route to a successful future is abundantly clear. The private sector is the engine of economic growth—so turn it loose. Global economic integration is the engine of development—so break down the barriers to free trade. Individual initiative and expanding prosperity are the best ways to improve human welfare—so provide opportunities and incentives.

These beliefs have long been held in one form or another by Western business leaders. But with capitalism now the world's reigning ide-

ology, such views are gaining wider acceptance. Indeed, there sometimes seems little difference in the economic policies espoused by the
Democratic and Republican Parties, by Great Britain's Labour Party
and Tories, even by Western governments and the Chinese Communists. Imagine an optimistic vision of the next half century consistent
with these views, a future in which freeing market forces and expanding the global market bring prosperity and social progress to a larger
and larger share of humanity. Call this vision *Market World.*[2]

*What had begun tentatively toward the end of the old century
exploded in the new. Country after country adopted the formula:
privatize, deregulate, rein in public spending, and unleash competitive market forces. Join the global market by dropping tariffs,
promoting exports, and seeking foreign investment. Build up
financial capital by encouraging savings and entrepreneurship.
Build up human capital by emphasizing education and health.
Create a modern infrastructure for transport and communications. Modernize the state by containing corruption, cutting red
tape, and revamping the legal framework to protect property
rights and facilitate commercial transactions.*

*That strategy, pioneered by the "Asian tigers," exported well.[3]
First, China made it work; then, Chile and other Latin American
countries; then, Uganda and a clutch of African countries. Russia,
once its political situation stabilized, also proved adept at turning
its educated workforce and extensive natural resources into an
economic powerhouse. The transition took longer in India, with
its massive central government, and in the xenophobic Middle
East. But India eventually grew rapidly, its well-educated middle class giving it a huge advantage, and Iran and a few other
Middle Eastern countries found ways to preserve a conservative
Islamic culture while joining the global market.*

*The result was a worldwide economic boom of unprecedented
breadth and longevity. By 2020, most countries of the world were
integrated into the global market. The Free Trade Zone of the
Americas; the South Asian Free Trade Zone; the European Union,
now including Russia and most of central Europe; and the African*

*Common Economic Market—all were thriving. Financial integra-
tion was even further along, marked by a single global financial
market with daily volumes measured in tens of trillions of dollars,
larger than the annual output of any single country. Now, there is
even talk of a global currency, modeled after Europe's euro. By its
sheer size and speed, the global financial market became the de
facto regulator of national economic performance as investors bid
down the value of currencies, upped interest rates, and reduced
the flow of investments to countries perceived to be less competitive.*

*With the global market came other changes. Nations that once
competed for territorial dominance now compete for market
share, with money that once supported military forces invested in
new ports, roads, telecommunications, and other infrastructure.*

*Part of what propelled the global boom was the spread and
continuing refinement of information technology. By 2020, Earth
was connected by an extensive network of fibers, satellites, and
cellular telephone links that not only tied the global market
together but also enabled new forms of commerce and new patterns of development to emerge. Who would have guessed that Sri
Lanka would become a key financial center, the Switzerland of
Asia, managing huge portfolios and investment services; that
African craft cooperatives would market their carvings and blan-
kets worldwide; that remote gambling and game services would
emerge as such a huge global business? At the same time, other technologies—biotechnology, molecular engineering, new energy
technologies—have proved important as well, generating succes-
sive waves of innovation and economic expansion.*

*But two other factors have also played a critical role in this
new era of prosperity. Large transnational corporations became
conduits of technology and expertise and sources of patient investment capital that helped jump-start economic growth in many
developing regions. And a proliferating host of voluntary international organizations, large and small, served as the mechanics and
maintenance workers of the global market, making rules, setting
standards, resolving trade conflicts.*

With global economic integration has come an increasing convergence in consumption patterns and in production methods. Advertising has in part driven the first; competitive pressure and the rapid spread of new technology, the second.

It's true that the leaders of a number of countries have foolishly tried to ignore the global market—and those countries have paid an enormous price. A few are backwaters yet, living on the raw materials they can sell and on the money their expatriate workers send home; others are making progress but are decades behind. But there is little sympathy or aid for such places. The choice to succeed or not is theirs.

Some decry the materialism of this global society; others, the fact that huge inequities, even a small amount of abject poverty, persist. But they overlook what has been achieved—a sixfold expansion of economic activity over the past half century and unprecedented prosperity, with a sizable middle class providing an anchor for social and political stability in most countries. The world's population peaked a decade ago, in 2040, never having reached 9 billion, and has been declining ever since. There have been enormous advances in science and medicine—vigorous health at eighty and ninety years of age now prevails. Pollution has been sharply reduced almost everywhere; waste is seen as a sign of inefficiency and poor management. It's not that all environmental problems have disappeared: the climate is changing, however modern societies can easily adapt to the changes. But everywhere the power of human ingenuity and enterprise is evident. Many great universities and research institutions have arisen, endowed in part by private benefactors, and numerous foundations underwrite scholarships and cultural endeavors. The arts, sports, and virtually every form of entertainment are flourishing— testimony to the wealth and leisure time prosperity affords.

Market World may sound optimistic, but achieving such a future requires only an acceleration of the globalization and economic reform already under way. How plausible is this scenario?

Those who are enthusiastic about the potential of *Market World*

point to the decades-long economic dynamism of Asia—now interrupt-
ed, but likely to resume—to an awakened and vigorously growing India
and Latin America, to the possibility of a turnaround in parts of Africa.
In the past few years, the global economy has been growing almost
twice as rapidly as in the preceding two decades, and some analysts are
extremely optimistic about the long-term prospects. If a major war or
an environmental catastrophe doesn't intervene, as Harvard University
economist Jeffrey Sachs told the *Wall Street Journal,* "economic growth
will raise the living standards of more people in more parts of the world
than at any time in history."[4]

Globalization is also surging ahead, with international trade growing even more rapidly than national economies. Indeed, by 2005, international trade is expected to account for 40 percent of national output
in industrial countries and more than 50 percent in developing countries.[5] International financial flows, only $20 billion per day fifteen
years ago, are now more than $1 trillion per day; by 2015, they may
reach $30 trillion per day.[6] For better or for worse, the trend toward
global economic integration has enormous momentum.

At the same time, the spread of capitalism may further the spread of
democracy and freedom. The very nature of economic activity in free
markets, many argue, requires broad access to information, the spread
of competence, and the exercise of individual decision making throughout the workforce—conditions that are more compatible with free societies and democratic forms of government than with authoritarian
regimes. And with greater wealth and more responsive governments,
societies will be both more able and more likely to solve environmental
and social problems.

But will economic booms and global markets benefit most people
and most countries, or only a few? On that, the evidence is less promising. Inequalities in wealth and income are rising sharply in Russia, eastern Europe, and China as market economies take hold in these regions,
and the gap between rich and poor nations is widening, not shrinking.
So believers in a *Market World* have to take on faith that eventually
prosperity will spread. When economic growth is still elusive in more
than a third of the world's countries, when squalor and corruption
remain widespread in developing regions, when nearly 2 billion people
lack access to such basics as a toilet, there is still a long way to go.

Nonetheless, many economic trends support the plausibility of a full-fledged *Market World*. But there is other evidence, too, from the pace and breadth of technological innovation, which some say could fuel a decades-long period of rapid growth as well as help provide new solutions to social and environmental problems.

Technological Transitions

Deep in the forest of the Brazilian Amazon lies the town of Japurá, the site where Chico Mendes, leader of the Brazilian rubber tappers, was brutally assassinated in 1988. A colleague of mine had reason to visit the rubber tappers there in 1994. Far from civilization, Japurá is reached by flying for several hours over unbroken forests and then circling the "airfield" until boys race out and herd away the cattle that normally graze there. To reach a portion of the forest set aside as an "extractive reserve" for rubber tappers and a neighboring "indigenous reserve" for native forest tribes requires a further six-hour trip by small boat.

Until the past decade or so, the forest dwellers in this region, protected by the vastness of the Amazon forest, had a lifestyle that had changed little for thousands of years. But what my colleague saw while touring the extractive reserve astounded him: a group of Indians, in the midst of this remote forest, carrying some of the latest high-tech tools of the twentieth century—a handheld global positioning system (GPS) device and a cellular telephone. It turned out they were demarcating the boundaries of their reserve with the GPS device—which uses U.S. military satellite signals to locate position very precisely—and using the cellular phone to call tribal headquarters for reinforcements whenever they found evidence of illegal logging on their reserve. In effect, they were protecting their forest and their pastoral way of life with the aid of modern information technology.

I've repeated this story here because it symbolizes for me how small a place today's world really is—and how large is the potential for change, especially when new technologies can change lives by giving people the ability to help themselves.

Less dramatically but just as surely, personal computers, cellular phones, and the Internet are altering the industrial world as well. In fact, the explosion of information technologies is just the most visible example of our era's extraordinary burst of technological change. That change, say historians of technology, comes in two ways: gradual improvement of a technology or a family of technologies through a series of small improvements; and more radical innovations that create very different ways of meeting human needs or permit totally new types of human activity. Although some economists and economic historians argue that it is radical technological innovation that drives major gains in human prosperity and human welfare, the incremental type of innovation is important, too.

Gradual improvements in energy technologies have increased the efficiency of energy use by about 1 percent per year—a century-long pattern—halving every seventy years the amount of energy needed to accomplish a given task.[7] An even more striking example comes from the silicon chips at the heart of personal computers. In 1970, Gordon Moore, one of the founders of Intel Corporation, noted that improving semiconductor technology doubled the number of transistors or other circuit elements contained on a single chip about every eighteen months to two years—a pattern since dubbed Moore's Law. The pattern has held for more than twenty-five years, producing astonishing gains in memory capacity and computational power, and seems likely to continue until at least 2010. By that time, standard personal computers are likely to operate at 10 billion cycles per second (100,000 times as fast as Intel's first processor in 1971 and about 100 times as fast as typical computers bought in mid-1996).[8] This pace of sustained innovation—and a nearly comparable one in telecommunications—is so rapid as to constitute an industrial revolution, with what are likely to be profound economic and social effects.

Still another example—this one with profound implications for whether the world can feed itself—comes from improvements in agricultural technology. Yields of grains worldwide have increased by about 2 percent per year over the past fifty years, doubling every thirty-five years the amount of food that can be grown on a plot of ground. In some countries, yields have grown even more rapidly: Egypt, for example, doubled its wheat yields in twenty years. Despite being dubbed the "green revolution," agricultural advances are a classic example of incremental technological improvement in new crop varieties, pest control, and improved farming techniques. But by whatever label, these improvements in yields have forestalled the famines predicted by neo-Malthusian prophets a generation ago. They allowed India, for example, to transform itself from a grain importer to a grain exporter, even while doubling its population. Agricultural applications of biotechnology, still in their infancy, may help spur the pace of innovation.

Although there are theoretical limits to energy efficiency, to how small electronic circuits can be, and to agricultural yields, there seems to be no fundamental reason why at least energy efficiency and agricultural trends cannot continue for decades to come. If such trends do continue, then by the year 2050, industrial societies might use 40 percent less energy (and produce 40 percent less pollution) than they otherwise would have, without any major shifts in energy policy or deliberate changes in lifestyle. Likewise, the world as a whole might be able to grow 2.7 times as much food on the amount of land under cultivation today—easily enough, in theory, to feed an expected population 50 percent larger than today's. Of course, social and environmental factors addressed in other scenarios complicate the picture.[9]

Incremental innovation is important, but there is no evidence that it generates an extended period of rapid economic growth and prosperity. Historically, such booms *are* associated with periods of more radical technological innovation. Some experts say that radically new inventions tend to occur in waves or clusters. According to one recent study, three such periods of intense creativity have occurred in the United States over the past two centuries.[10] The country—and the world—may now be in the early phases of a fourth such period, one representing the most profound cluster of radical innovations since the harnessing of steam energy in eighteenth-century England. Not only is an information revolution in full swing, but so also is a biotechnology revolution, involving manipulation of the chemistry of life itself. Scientists can now construct materials almost atom by atom or molecule by molecule, giving rise to a nascent revolution in high-tech materials, molecular engineering, and very small devices (so-called nanostructures). And there is a new generation of radical energy technologies—fuel cells and solar cells—on the horizon. So the technological fuel for the extended boom described in the *Market World* scenario is at least plausible.

Besides transforming industrial patterns, radical technological change often seems to be accompanied by a burst of social and institutional creativity. For example, the automobile industry, sparked by the inventions of the internal combustion engine and petroleum refining (to produce gasoline), could not have become such a major factor in the twentieth century without the development of mass manufacturing techniques by Henry Ford and others and the creation of new forms of finance, such as sale of corporate stocks.

The period between 1880 and 1915, sometimes described as a golden age of capitalism, witnessed not only the beginnings of the automobile and petroleum industries but also the invention of Thomas Edison's lightbulb, the electric dynamo or generator, and the telephone, initiating the electric power and telecommunications industries. The first plastic was invented, enormously expanding the chemical industry, and the aviation industry sprouted its wings following the development of the Wright brothers' first plane. Economic growth for the period averaged about 2 percent, low by today's standards but more than double the rate of the preceding half century.[11]

Is the present a comparable period, and will it also initiate a sustained, perhaps worldwide, period of prosperity? A historical comparison may be useful to provide a sense of just how radical a change the revolution in information technologies represents.[12] It was the steam engine that made possible the industrial revolution, changing dramatically—by a factor of about 1,000 over a few decades—the amount of power at humanity's disposal. For the information revolution, the comparable technologies are the ability to manipulate information with computers and to transmit it over an optical fiber or some other telecommunications link. And computing power and telecommunications bandwidth (the rate at which information can be sent down a single communications channel such as an optical fiber) are changing even more dramatically, increasing by factors of 100,000 and 10,000, respectively, in just three decades. So the information revolution appears to arise from a far more powerful technological transformation than the one that drove the industrial revolution.[13] Prices per computation or per bit of information transmitted have already fallen dramatically and are still falling. Moreover, these new technologies are far more accessible to ordinary individuals than were the initial fruits of the industrial revolution. And they are spreading globally with remarkable speed; Intel reports, for example, that in 1996, nearly as many new Pentium computers were sold in Asia as in the United States.

Just as in other waves of radical innovation, information technology is transforming the productive process itself—permitting just-in-time manufacturing, for example—and altering corporate structures as companies begin to make use of the ability to assemble, organize, analyze, and communicate huge volumes of information. Moreover, new forms of finance, such as venture capital, have emerged along with other forms of social innovation, from "virtual" corporations to on-line chat groups.[14] The information revolution is making possible truly global markets and companies and is transforming warfare with "smart" weapons, sophisticated sensors, and theaterwide communications. But it is important to realize that most of the economic, social, and—probably—political effects of the information revolution are still to come.

Nearly twenty years after the introduction of personal computers, less than 50 percent of U.S. households own one; in Europe, the figure is less than 20 percent and in Japan, less than 10 percent. Fast digital communications links will not reach most U.S. households until the early years of the twenty-first century. Despite the explosive growth of the Internet, use of which is doubling about twice per year, less than 25 percent of the U.S. population, and less than 1 percent of the world's population, is on-line.

So the interesting question is whether the information revolution and the other emerging radical innovations will usher in a new era of rapid economic growth. The historical precedents described earlier make such an era seem plausible. But the information revolution is also likely to affect the spread of knowledge, the nature of work, and, possibly, the forms of governance as on-line commerce, libraries, and government services grow from fledgling experiments to commonplace institutions.[15]

Moreover, the information revolution will spread globally far more quickly than did the industrial revolution—to newly industrializing countries such as China and even to preindustrial societies such as the Amazon forest dwellers near Japurá. It is possible to foresee a time, not many decades hence, when nearly every village on the planet will be connected to information networks via global wireless and satellite links.[16]

Is the vision of global market growth, of a far more prosperous *Market World*, realistic? Yes, and for far deeper reasons than just a faith in free markets. Powerful economic and technological trends all point toward a period of unparalleled prosperity. But such a future will not necessarily come to pass, and even if it does, it may not bring prosperity to all.

Free markets and rapid economic growth by themselves may not be enough to create a more optimistic future. For one thing, markets can't do everything. "Capitalism is myopic and cannot make the long-term social investments in education, infrastructure, and research and development that it needs for its own future survival" is how the distinguished economist Lester Thurow describes some of these limits.[17]

A largely market-driven future, one devoid of major and deliberate interventions to shape the course of human destiny, includes a number of risks. More specifically, it includes environmental risks—that the price of prosperity could be a degraded and permanently altered Earth for future generations; equity risks—that the benefits of growth could continue to accrue unequally, leading, some would say, to "a one-sided dictatorship by the rich and the powerful, or unbridled and unfair competition between the rich and the poor;"[18] social risks—that growth could come too slowly in poor countries largely outside the global market, leading to greater poverty and perhaps social disintegration or massive migrations; and security risks—that those denied a share of the new wealth and a chance at a better life could eventually seek redress violently where political channels are ineffective or closed to them.

In short, a *Market World* scenario risks unattended environmental problems that could darken the future for all and unattended social problems that could deny the future altogether to some countries and regions.

Chapter 4

Fortress World: Instability and Violence?

"Islands of prosperity, oceans of poverty."
—*Madhav Gadgil*[1]

IF *MARKET WORLD* FAILS—if global markets and rapid economic growth do not lift the bulk of humanity out of poverty or if escalating environmental change, perhaps even widespread ecological collapse, and growing social stresses undermine prosperity—might civilization itself unravel to some degree? Might the world or large portions of it descend into a future characterized by cruelty and chaos and fear, by unprecedented human misery? Perhaps most forbidding, might such a future see widening gaps and bitter conflict between rich and poor segments of society and between rich and poor nations, barring any hope of global cooperation? Would the wealthy be able to protect them-

37

selves, or might they, too, find their future dimmed? Imagine such a world divided against itself, armed and guarded—a *Fortress World.*[2]

The rapid worldwide economic expansion that characterized the first two decades of the new century brought prosperity to perhaps only a third of humanity. As the power of global markets and the prominence of global corporations grew, intensifying economic competition for investment and export markets, governments in many developing countries downsized, sold off state-owned companies, and loosened regulatory controls. Attention to social and environmental concerns declined. But the boom bypassed whole regions, and even in rapidly growing countries the surge in income and wealth was highly concentrated.

In many rural areas, incomes dropped and living conditions deteriorated as the century progressed. Although productive land remained plentiful in Latin America and parts of India, much of it was locked up in large estates; Africa's burgeoning population faced acute shortages of farmland. Landless farmers, and the sons of those whose tiny plots could not be divided further, increasingly pushed into marginal lands, cultivating hillsides and clearing forests. But marginal land eroded quickly, and even good land degraded when poor farmers failed to renew soils or grazed too many cattle on shrinking grasslands. The result was increasing rural poverty, an unprecedented flood of urban migrants—people no longer able to make a living from the land—and the rise of vast urban shantytowns, dwarfing those of the 1990s.

In contrast, incomes in the industrial world and in middle-class enclaves in developing countries continued to rise, at least for a while, and lifestyles expanded apace. Advertising in the global media and a booming travel industry brought new awareness of this affluence and fueled growing resentment among the wider masses of people, who desired the comforts and consumer products of the rich but could not hope to achieve them. Teenagers—now more than 1 billion in number worldwide and a volatile, mobile group—were especially susceptible to these influences. In every developing city, teenagers not only staffed the informal economy as street vendors and domestic servants but also joined idealistic and nihilistic groups of every description, including a rising number of terrorist organizations—a growing army of the angry poor.

Economic expansion also brought rapidly worsening pollution to much of industrializing Asia and Latin America as sprawling new factories spewed wastes into the environment. Ever larger numbers of cars and trucks choked urban streets, making the air hard to breathe. Deteriorating health conditions were increasingly evident—chronic lung disease in urban areas, an epidemic of cancer from polluted waterways, virulent new diseases emerging from devastated forests and estuaries. But lack of money, political will, and often the technical ability to enforce environmental laws meant that governments in most newly industrializing countries did little to rein in pollution. Outside help for environmental and social problems declined, too, as foreign aid—widely unpopular in rich countries—dried up.

There were broader signs of environmental distress. One by one, the major marine fisheries collapsed, victims of sustained overfishing by huge trawler fleets eager to supply international fish markets. Fish, produced almost entirely by aquaculture now, became a luxury food. Fishermen lost their jobs, but more devastating was the loss of the primary source of protein for three-quarters of a billion people. Wood, too, became scarce and valuable as forests disappeared, and hundreds of millions of people lacked the firewood to cook their dinners. The reality of climate change was no longer debated—drought cycles were unquestionably more intense and more frequent than in the past, and severe flooding devastated large areas of farmland and low-lying coastal areas.

As conditions became increasingly desperate, the voices of the disenfranchised got louder. In India, what started as a protest march by a group of fishermen became an army of more than 2 million poor people that converged on New Delhi, demanding food and redress of their dire economic straits. In Mexico City, an unusually vicious killer smog left thousands of Mexicans dead and hundreds of thousands ill, igniting a protest in which people shut down the highways, bringing the city to a standstill. Small-scale conflicts over scarce grazing lands, over access to shrinking forests and woodlots, and over the allocation of irrigation water became increasingly common.

Some of the conflicts became extremely violent. Syria and Turkey fought a short but deadly war over the water of the Euphrates River. Attempts by landless farmers in Brazil to take over idle estates turned into bloody massacres at the hands of the landlords' hired armies. During an economic downturn in China, protests by laid-off construction laborers in several major cities turned into food riots and looting, with thousands of deaths when army troops opened fire; the conflict nearly brought down the government.

Contributing to the upsurge in violence, felt even in the industrial countries, was the growing power of organized crime on a global scale. Criminal organization now controlled governments in several developing countries, gaining a secure base for their trafficking in drugs, weapons, and illegal migrants. Increasingly elaborate computer crimes defrauded millions and toppled more than one major international bank. Tacit alliances between criminal organizations and local terrorist groups posed a worrisome security threat: a heroin-for-nuclear-warhead barter between the Russian mafia and a Pakistani terrorist group intent on detonating the device in a U.S. city was intercepted only by accident.

In the rich countries and in well-to-do-enclaves in the developing world, armed clashes among competing criminal groups and repeated terrorist incidents created a growing sense of vulnerability. Public and private spending on security rose sharply, walled and guarded communities became a way of life, and many business executives employed bodyguards for themselves and their families. Few tourists visiting developing countries ventured beyond officially sanctioned sites and luxury retreats, all of which were carefully guarded. Wealthy farmers, ranchers, and other large landowners hired army patrols. Trade suffered and international investment slowed.

A flood tide of illegal migrants poured into rich countries, adding to the countries' sense of being under siege. Immigrants were blamed for crime, for unemployment, and for the spread of new diseases. The political demand to do something, anything, about illegal immigrants became overwhelming. Europe repealed the Common Market's open border provisions, required everyone to carry a high-tech identification card, and instituted random street and highway checks; any foreigner lacking proper identification was immediately deported. The United States, too, issued identity cards, removed due process provisions for noncitizens, and built an elaborate set of fences, concrete barriers, and electronic sensors along its southern border. Nonetheless, racial and ethnic tensions escalated.

Africa's collapse, when it came, was not entirely a surprise. Despite attempts at reform, the continent's rapidly rising populations, falling incomes, and corrupt and self-serving governments proved overwhelming. In the decaying cities, crime was often the only way to feed a family. The collapse began in West Africa as state after state foundered, but over the next few years it spread across central Africa and then into East Africa, leaving near anarchy in its wake. Huge numbers of desperate refugees overwhelmed the few stable countries and threatened southern Africa as well. At the height of the troubles, more than 5 million people per year died of violence, hunger, and disease.

Civil order collapsed in other regions, too, if more sporadically. The disparities between rich and poor in Latin America reached extraordinary levels, as did resentment; rural struggles over land and urban uprisings against police violence often unleashed widespread insurrections. In China and India, massive unemployment created explosive social and political conditions. In North Africa and the Middle East, growing populations and social problems swelled the ranks of radical Islamic factions and led to violent revolution against autocratic governments, but the upheavals did little to improve the underlying problems.

Faced with chronic instability in developing regions, the industrial world turned inward and the world economy stagnated. In their protected enclaves, their Fortress Worlds, the rich were like islands of prosperity in an ocean of poverty. And yet even there, they could not entirely escape. The poor, bereft of all else, still found ways to export their misery in the form of crime, violence, and disease.

This scenario sketches, in deliberately dramatic form, a vision of what might happen if *Market World* fails. Is this a plausible scenario? Existing trends are certainly suggestive.[3]

Although private investments in developing regions are rising—fueling China's spectacular growth, for example—less than two dozen countries now benefit to any significant degree. In more than seventy countries, incomes are lower now than they were in 1980; in forty-three countries, they are lower than in 1970.[4] Most developing countries still qualify as "roadkill on the global investment highway," in a financial journalist's heartless phrase;[5] a large proportion of the world's nations are in economic decline; and more are growing only slowly or sporadically. Evidently, the global market boom is still far from truly global.

Even where economic growth is rapid, problems remain. In their haste for growth, most newly industrializing countries ignore environmental concerns. As a Chinese aphorism puts it, one should not "give up eating for fear of choking"—a view cited even in an official report of the Chinese Environmental Protection Agency. As a result, many of today's burgeoning cities in Asia and Latin America are already gasping for breath. The huge smoke cloud that shrouded much of Southeast Asia for months in 1997, causing widespread respiratory disease and closing schools and airports, is a dramatic illustration. More insidiously, health experts talk about the growing toxic legacy from uncontrolled industrial pollution that may soon burden public health in China, India, and other rapidly industrializing countries much as it already does in eastern Europe and Russia.

Given the rapid growth, until recently, of prosperity in much of Southeast Asia and now China, why are so many countries being left by the wayside? One reason is that, as the *Economist* puts it, market-based economies require "a complex web of effective institutions, from basic property rights and well-run legal systems to effective and incorrupt bureaucracies. In poor countries such institutions are often weak or nonexistent."[6] Moreover, creating such institutions where they don't exist is not easy; it requires willing and well-informed leaders, a broad base of political support within a country, and—if outsiders are to help—a more sophisticated type of intervention. All too often, in the absence of effective institutions, large amounts of foreign aid funneled through governments simply help to corrupt and enrich the officials of those governments—exemplified in the extreme by the lavish French villas and Swiss bank accounts of Zaire's deposed ruler Mobutu Sésé Séko.

Contrasting examples from India illustrate the link between economic growth and stable social institutions. The "green revolution" agricultural reforms that enabled the country to more than double its grain production between 1970 and 1990 took root only in several Indian states, most notably in the Punjab. There, farmers, supported by a capable local agricultural university and state efforts to make credit available, could borrow money to buy new seeds and fertilizer. An existing network of roads made it possible for the farmers to get their crops to market. As yields rose and farmers' income grew, the techniques spread, farmers borrowed more, and yields rose still further—and the threat of widespread famine in India disappeared. But the new techniques have never been adopted in what is potentially the most fertile part of India, the huge Gangetic Plain, which could, as one agricultural expert put it, "potentially feed much of Asia."[7] This region remains one of the poorest in the country, with low literacy and poorly functioning state governments; it lacks roads, credit institutions, marketing cooperatives, an effective agricultural university—in effect, a whole web of social institutions and infrastructure needed to make market-based agricultural growth possible.

The contrast between the Punjab, one of the wealthier states in India, and the Gangetic Plain, or the contrast between rapidly industrializing China and stagnating states such as Angola (which, ironically, is rich in oil and minerals and has a higher per capita income at present than does China) and Colombia (beset by endemic corruption and cocaine-related violence), illustrates the dilemma. Without social and institutional development, economic growth either doesn't happen or remains sporadic and the global market remains largely irrelevant. And without economic growth, poverty doesn't go away. In such "nondeveloping" countries, the risks of social instability and conflict can but rise.

Examples of such conflicts are already appearing—in Latin America, in Africa, in India—and they are likely to intensify in coming decades as social and environmental stresses rise. Most of sub-Saharan Africa, for example, cannot feed its population adequately now, yet the region's population is expected to triple in the next half century. India, too, faces growing rural impoverishment from the combined effect of expanding population, environmental degradation, and diversion of resources to urban and industrial uses.[8] Water, already scarce in North Africa and most of the Middle East, will become more so if the population doubles as expected; and much of this scarce water must be shared among unlikely partners—Turkey, Syria, and Iraq; Israel and its Arab neighbors—adding to tensions in an already conflict-prone region. "There can be no enduring peace in the Middle East without water-sharing arrangements," says water expert Peter Gleick.[9] And when oil production and revenues decline, as they will over the next two decades for all but a few countries in the region, social pressures may well become explosive.

Under such conditions, many poor people will try anything to get a chance at a better life. Illegal migration is already a serious problem, with what historian Paul Kennedy calls "demographic fault lines" separating the United States and Latin America, Europe and North Africa, Bangladesh and India, and China and Siberia[10]—in addition to tens of millions of internal African migrants from poor countries such as Mali to more prosperous ones such as Ivory Coast, Gabon, and South Africa. Smuggling such desperate cargo is already a lucrative business, with young Bangladeshis paying $5,000 apiece for illegal entry into Germany and some Chinese emigrants paying $20,000 for passage to the United States. Yet the income gap between have and have-not countries will get much larger.[11]

An even larger demographic shift is under way. Cities in developing countries will expand enormously, adding 2.5 billion people by the year 2025, mostly from rural areas. The concern is whether cities will absorb this flood tide of urban migrants or succumb to crime, pollution, and unemployment-driven instabilities. Global criminal groups are already outrunning national police efforts with sophisticated schemes to move illegal goods over borders and to hide their money in global financial markets. That money is a powerful corrupting force, partly because there is so much of it—an estimated $500 billion per year in the drug business, more than the gross national product of any developing country. Russia, Mexico, Colombia, Nigeria, Bosnia—the list of countries struggling with this new threat is growing. In response, corporations and individuals are increasingly hiring private security forces, which now outnumber public police officers by three to one in the United States; in more lawless places like Russia and South Africa, the ratio is ten to one.[12] Of even greater concern to national security agencies is the burgeoning global market for weapons, bomb-making information and materials, and agents of chemical and biological warfare. Even nuclear weapons and the components to make them are not secure; Aleksandr Lebed, the Russian general who became a politician, says bluntly that his country's efforts to keep nuclear warheads off the black market are "unsatisfactory."[13] If these dangerous technologies can be had for a price, how long will it be until terrorist groups succeed in acquiring them?

Environmental trends affect rich regions as well as poor ones, and it will take global action to head off climate change; rich regions cannot by themselves alter this trend. Nor can they entirely protect themselves from virulent new diseases or the northward migration of existing ones as the climate warms.

The *Fortress World* scenario presents a dark vision of the future. It assumes that the capacity for cruelty and violence and for apathy regarding the suffering of others are enduring human traits. It asserts that the failure of market-based growth to achieve widespread gains in human welfare could undermine the social contract that binds societies together. It foresees the possibility of economic stagnation, with many developing markets in disarray; of widespread inequity exacerbated by illegal seizures of land, water, oil, and other resources; of the breakdown of social order; and of widespread instability. In such a divided world, the scenario asserts, conflict between rich and poor would be endemic, and cooperation on global issues—or any kind of protection for endangered wildlife and rare habitats in desperate developing countries—would be unlikely.

The result, the scenario argues, would be more crime and conflict, reinforcing the fortress mentality of the rich, exemplified by Moscow's modern capitalists, who live and work guarded by armed men for fear of the violent Russia mafia. It is not a world most people would choose to live in.

Need our future be so grim?

Chapter 5

Transformed World: Changing the Human Endeavor?

"Never doubt that a small group of thoughtful, committed citizens can change the world; indeed, it is the only thing that ever has."
—*Margaret Mead*[1]

IS THERE AN ALTERNATIVE to the single-minded pursuit of prosperity that is *Market World* or the descent into chaos and cruelty of *Fortress World?* Imagine a society that seeks not just wealth but also human welfare, not just security but also fairness. A society that is a steward, not an exploiter, of Earth. I have chosen to call such a future *Transformed World* to emphasize the possibility of fundamental change for the better—in our politics and policies, in our social institutions, in our environment, in our collective behavior and cultural norms. Such a

47

future is not assured, but, as we will see, many trends already evident suggest what might be possible.

> *Preliminary signs of fundamental change emerged in the last decade of the old century. Smoking, once widely practiced in the United States, came to be regarded as a loathsome habit and was banned in workplaces and public buildings, with tobacco companies widely denounced as socially irresponsible organizations. Who would have imagined that such powerful corporations would be restricted in what they could sell and how they could sell it and forced to pay huge sums for the health damage they had caused? Or take the uproar in Europe that greeted the Shell Oil Company's decision to discard an obsolete drilling platform on the bottom of the North Sea. A campaign orchestrated by the grassroots environmental group Greenpeace forced this huge international company not only to abandon its plan but also to rethink its corporate strategy. At the same time, a few governments became zealous defenders of the environment. In what were later seen as precedent-setting moves, both the Netherlands and New Zealand adopted visionary "green" plans to remake their societies within a generation.*
>
> *The new century saw a radical turnaround in environmental thinking as evidence of a shift in Earth's climate accumulated: in the United States alone, nearly half a dozen "one-hundred-year" floods devastated communities along the Mississippi River; hurricanes of unprecedented force repeatedly struck the East Coast; back-to-back years of record heat waves hit a number of major cities, killing hundreds. Public opinion shifted in favor of taking action. In a widely quoted speech before the United Nations, the president of the United States harked back to World War II rationing, called for a spirit of shared sacrifice, and proposed a drastic reduction in emissions of greenhouse gases. What eventually emerged, despite fierce opposition from the coal industry, was an agreement to cut emissions in half over a thirty-year period by rationing ever more tightly the types of fuels that cause greenhouse gases—in effect gradually but inexorably raising the price of polluting fuels.[2] A binding international treaty signed by most*

countries required them to adopt equivalent measures, with less stringent targets for developing countries, reflecting their lower emissions.

The results were dramatic. Within ten years, energy use in U.S. industries dropped by a quarter and gas-guzzling vehicles became as socially unacceptable as smoking; in ten more years, U.S. emissions had nearly reached levels specified in the treaty and European cuts had gone even further. New cars that averaged nearly eighty miles per gallon were so popular that manufacturers could not keep up with demand; and with costs for conventional fuels high, alternative sources of energy took off: wind and solar energy became big business. New and more efficient energy technologies continued to appear, and by 2050, there were signs that the climate had begun to stabilize.

At the same time, as governments used their revenue windfalls from energy-related taxes and permit sales to lower social security and other employment taxes, labor costs came down and employment surged, especially in Europe. The climate treaty also brought a flood of energy-related investments in developing countries, resulting in huge gains in energy efficiency and reductions in urban air pollution.

Lower taxes proved so popular, and the logic of reducing pollution by boosting industrial efficiency so compelling, that many countries also reduced income taxes and raised their levies on other natural resources in addition to energy, taxing such raw materials as metals, minerals, forest products, and even water. That extended the revolution in industrial efficiency: recycling rates for many materials soared. Europe's industrial "take-back" laws, which required manufacturers to reclaim and recycle most durable goods at the end of their useful life, were widely copied. At the same time, there was a pronounced decline in the consumer culture of the well-to-do countries, especially among younger people: "low-impact" lifestyles, vegetarian diets, and antimaterialist ethics gained a wide following. With less need for raw materials and energy, pollution and waste-disposal problems in the industrial countries declined sharply.

But in parallel with these sweeping environmental improve-

ments came equally dramatic social changes, such as those that
occurred in America's inner cities and among its underclass.
Called the "urban renaissance," what began as welfare reform
gradually became a more sweeping transformation. City after city
set out to reduce crime, revive its inner core, create jobs, and
reduce poverty, drug use, and other social problems. Radical
efforts to improve public education and to create parks, green-
belts, and other environmental amenities became commonplace.

Equally important, however, was the religious revival that
filled churches and even athletic stadiums, emphasizing family ties
and demanding personal commitments to higher standards of
behavior. Following the lead of many black churches, religious
groups of every denomination enlarged their social ministry. As
evidence of these groups' effectiveness accumulated, many cities
formed partnerships with them and other community groups, providing them with funds to deliver social services, support families
under stress, and motivate job seekers and recovering alcoholics.
Within a generation, poverty had declined and virtually every
social indicator showed marked improvement. Although experts
debated whether public programs, religious groups, or simply a
strong economy had contributed the most, the progress was unde-
niable. And with urban revitalization, cities again became attractive, exciting places to live, greatly slowing suburban sprawl.

The involvement of religious and community groups in America's urban renaissance was part of a larger phenomenon of the
new century: a huge increase in most parts of the world in the
number and influence of citizen's groups, voluntary organizations,
and other so-called nongovernmental organizations. Devoted to a
host of environmental and social causes, they included local community groups, large activist organizations such as the Sierra Club
and Greenpeace, traditional charities and emergency relief groups
such as the Red Cross and CARE, watchdog groups modeled on
Amnesty International, international development organizations,
independent think tanks, and religious congregations. They delivered services, lobbied governments, confronted corporations, and

alerted the news media—in effect, they fomented, facilitated, and sometimes demanded social change.

The impact of this citizen power came in part from the sheer numbers of these groups—more than 1 million at the beginning of the century and nearly ten times that by 2050—which far outnumbered governments and even corporations: such a multitude of ears, eyes, and voices transformed political activity. But even more, their power came from their ability, despite the bewildering array of causes they espoused, to form spontaneous coalitions around particular issues and to motivate and arouse public opinion.

The continuing spread of information technology helped to facilitate this quiet, bottom-up social revolution, in effect forging a unique tool kit that enhanced the effectiveness of citizen's groups by linking them together. Increasingly, these coalitions were national or even global in scope, with hundreds and even thousands of groups working together through the Internet and its successor, the GlobalNet. The successful campaign to ban child labor, for example, began with a network of community groups in India but soon found allies among social activists, labor unions, and religious bodies in the rich countries. Together, they formulated the Children's Rights Law and focused so much public attention on the issue that virtually all governments adopted it. In another example, the fact that some tropical forests survive today is largely attributed to the now famous Save the Forest coalition, which monitors logging activities in every country and regularly distributes its often devastating videos over the GlobalNet. Over the past half century, in fact, citizen's organizations have become a major agent of social change.[3]

Another equally remarkable change was the "greening" of the private sector. Despite their awesome economic power, global corporations—and many smaller ones, too—now set their strategies and conduct their activities in ways that take social expectations into account. At first, there were only a few companies committed to environmental and social goals as well as to profits. But as these "leader" companies broke ranks and publicly supported energy rationing to protect the climate or promote other emerging social goals, they provided an opening for citizen's groups— indeed, often worked directly with them—to bring pressure to bear on laggard companies. In industry after industry, codes of conduct incorporating social and environmental expectations were devised by coalitions of citizen's groups and industry representatives. Modeled after the Sullivan Principles that decades earlier had defined acceptable corporate practices in South Africa under apartheid, the codes banned sweatshops in the global garment industry, forced an end to most clear-cutting in natural forests, set standards for labeling produce as organically farmed and pesticide free, and made it unacceptable for manufacturers to pollute in poor countries in ways forbidden in rich ones.

For the most part, these standards were never codified in law, and compliance was voluntary, but their influence on corporate behavior was nonetheless profound. The ability of citizen's groups to monitor corporate activities in virtually all parts of the world and to publicize their findings on the GlobalNet and through other media proved to be as strong a force as legislation, and far more flexible. Companies that once bribed government officials with impunity found it impossible to bribe or silence an international network of hundreds of small organizations, and damaged reputations cost them far more in sales and, especially, in the ability to hire talented employees—their most critical resource—than would any conceivable fine. Financial markets, noting that companies in compliance with the voluntary standards usually fared better economically than did laggard companies, began to value companies accordingly, providing a powerful economic incentive. Over time, the result was a corporate culture that accepted the legitimacy of social goals and that worked with, rather than ignored, the communities in which they operated.

Yet another cause of the profound social change of the past fifty years was a surge in private philanthropy and a new activism on the part of foundations and wealthy individuals. Some scholars credit the activism to U.S. television magnate Ted Turner, who at the end of the past century gave $1 billion to the United

Nations, and financier George Soros, who gave nearly as much to support Russia and other countries in transition. Such philanthropists made it fashionable for the wealthy to compete in giving their money away rather than in accumulating it. In any event, much of the unprecedented wealth created by the late-twentieth- and early-twenty-first-century boom in technology gave rise to thousands of new foundations. The largest of these, the Gates Foundation, ultimately had assets of $50 billion, but even that staggering sum was dwarfed by the total of new philanthropic capital: more than $1 trillion in the United States alone by the year 2025. This money financed a large part of the social revolution of our time; it funded citizen's groups and international organizations, helped cities try radical experiments in education, supported free access to the GlobalNet for poor communities and villages, and assisted a host of other efforts aimed at creating a better world.

The GlobalNet itself was a potent catalyst for change. Its greatest impact came from facilitating widespread participation in social and political decision making—becoming in effect the nervous system of our global civilization—and enabling an enormous leap forward in developing regions by bringing a host of services directly to individuals and communities. It is hard to remember what life was like in such regions before every urban neighborhood and every village, no matter how remote, had access to educational programs, medical consultations, and agricultural advice; before village cooperatives and individuals could get loans, even for small amounts, via remote credit plans; before they were connected to the ferment of national and global events and the outpourings of a thousand cultures. The GlobalNet, by helping to fulfill the nearly insatiable human demand for access to information, exposed billions of individuals to wider horizons and new opportunities. Perspectives shifted in a single generation to a more modern social and political outlook, accelerating development.

Democratic forms of government and the rule of law have become almost universal, although they are still very new in some countries. Environmental conditions are gradually improving, and

every industrial nation and most developing nations now have
explicit social and environmental goals. Global population
reached a peak in 2040 and has since gradually declined, even
though Africa's population is still growing somewhat. There
remain great disparities in wealth and income, but these are
declining. For the first time, it is as a global civilization that
humanity conducts its affairs.

Many additional factors played a role in this achievement, of
course, such as growing prosperity and continued, indeed remark-
able, technological progress. A half century without any major
wars—initially because no nation could challenge the United
States militarily—made a big difference; as peace became the
norm, many countries followed the example of Central America
and disbanded their military forces entirely, using the resulting
"peace dividend" to finance social progress. Also contributing
were nearly universal literacy and a gradual rise in the status of
women. But the most important factor—almost entirely missed by
earlier forecasters—was the social and political revolution led by
citizen's groups that profoundly changed the nature of the human
endeavor.

How likely is such a scenario? Could bold political leadership and
new kinds of social activism and cooperation effect sweeping changes
in government policies and in the behavior of powerful corporations?
This vision of the future, sketched deliberately in optimistic tones,
requires a considerable leap of faith. But many events that have already
occurred give the central assumptions of this scenario a sense of real-
ism. Here are a few reasons why I think a *Transformed World* must be
considered a viable option for the future.

To start with, all of the twentieth-century examples mentioned in
the scenario are real. Shell and a growing number of other major corporations understand that meeting social expectations remains central
to their profitability and long-term success. Many companies, alone
and in industrywide groups, are voluntarily taking steps to lower their
environmental impact.[4] The British Petroleum Company has broken
ranks with other energy companies and publicly supported the need for

climate protection. A few business leaders are including environmental and social goals in their strategic business plans, remaking their companies as "sustainable enterprises."[5] Indeed, to further such goals, a group of 120 major global companies from thirty-five countries and more than twenty major industrial sectors have come together as the Geneva-based World Business Council for Sustainable Development, founded by Stephan Schmidheiny, a successful European entrepreneur.

Schmidheiny and others are convinced that such approaches are far more than just responsible corporate behavior; rather, they see them as critical to profitability and to competitive advantage in a global economy. Schmidheiny told me, for example, of his concern about the fact that forests are being cut down around the world. But his response is revealing. That means, he says, that forests will not be able to meet the rising global demand for wood and other forest products in coming decades—and hence prices will rise sharply. So he is buying forest properties and managing them "sustainably"—managing the forests to improve them and preserve them for the future, harvesting less than the annual growth—with the expectation that they will be far more valuable in the future.

To such companies, environmental groups are not enemies but allies. Indeed, once-unthinkable partnerships between environmental groups and individual companies or whole industries are devising novel approaches to pollution and waste problems—and then sometimes approaching government at the local or national level to get their approaches codified into law. McDonald's Corporation and the Environmental Defense Fund, for example, collaborated to find a way to recycle the plastic containers the chain's famous hamburgers are served in, the Procter & Gamble Company and the National Audubon Society have worked together on wildlife management, and a coalition of chemical companies and U.S. environmental groups negotiated an improved Superfund law to clean up toxic waste sites.

In contrast, companies and industries that flout social expectations are experiencing increasing difficulty. The current wave of litigation and regulatory action against tobacco companies, for example, has yet to end, but it seems likely that the outcome will fundamentally alter corporate behavior, and not just in the tobacco industry. Global companies are extremely powerful, but they can operate only with the sanction of society.

A few national governments are getting into the act, too, adopting strategic social and environmental plans in the same spirit as are the most innovative corporations—endeavoring to improve their societies and give their citizens long-term quality of life, even if it means sweeping changes. The Netherlands, for example, aims to transform the country's industrial base within a generation, creating a cleaner and more efficient economy and reducing pollution and waste. New Zealand plans to preserve its grasslands, forests, fisheries, rivers, and other natural assets while providing for greater equality in its multiracial society. Sweden has already begun to raise energy taxes (and hence reduce greenhouse gas emissions) and cut employment taxes—even though a warmer climate might ease its severe winters—because it is convinced that higher industrial efficiency and higher employment will pay off economically and socially.

The actions of a few countries, just like those of a few corporations, are not yet an overwhelming trend. But these instances show that long-term thinking and fundamental change can occur—even in corporations that file quarterly profit statements and democracies in which politicians face the pressure of upcoming elections.

At a smaller scale, the beginnings of an urban renaissance can already be seen in a number of U.S. cities. Chattanooga, Seattle, Cleveland, Detroit, and other cities are spearheading novel efforts in urban renewal and revitalization; at the neighborhood level, so are many church groups. In Arlington, Texas, for example, a mission housed in a building donated by a Baptist church and staffed by volunteers from fifty local churches provides a wide range of social services to more than 500 families per day. It is supported both by donations and by federal money passed on by the city government.[6]

Equally striking changes are under way in smaller communities and in rural areas.[7] In the arid region where Mexico, New Mexico, and Arizona come together, for example, a group of ranchers have banded together as the Malpai Borderlands Group to preserve their rangelands and their way of life. They have made common cause with ecologists to manage their lands for long-term preservation as well as profit and have given up their cherished self-reliance in favor of cooperative approaches such as a novel "grass bank" to help one another in times of drought.[8]

But if social transformations are possible at local and even national levels, might it still prove impossible to build a global consensus dealing with global problems, especially when such a step requires as many as 200 national governments to act in concert? The prospect sounds daunting, but already the world has agreed to three global environmental treaties that potentially commit countries to restructure their industrial base and to change the way they manage significant portions of their land. Only the first of these, the Montreal Protocol, which mandated elimination of stratospheric ozone-depleting gases, can so far be clearly judged a success; production of such gases have been phased out in the industrial countries, and the ozone layer itself is expected to recover by about the year 2050. Treaties to protect the world's climate and ecosystems have yet to be convincingly implemented, but last year's climate negotiations in Kyoto, Japan, brought that goal a step closer. What is distinctive about the most recent efforts is that they have been helped along and sometimes even led, informally but effectively, by coalitions of citizen's groups. The climate treaty, for example, was approved despite the reluctance of the governments of the United States and the former Soviet Union—the world's largest energy users—and Saudi Arabia, the world's largest oil producer, in large measure because of steady pressure from environmental groups in many countries.[9] More recently, a treaty to ban land mines was negotiated in less than two years, outside normal U.N. channels, and virtually all major governments except those of the United States, Russia, and China agreed to sign it—an effort led from start to finish by an international coalition of more than 700 citizen's groups. These groups, despite their informal style and hodgepodge of interests, nonetheless seem more efficient and nimble than governments or U.N. agencies and are becoming increasingly skilled at international diplomacy. And that augurs well for the eventual chances of a global approach to protecting the climate and for achieving other global goals.

To be sure, citizen's groups, with their bottom-up approach to social change—the ultimate expression of self-governance in free societies, as some commentators have described the phenomenon—may not succeed in catalyzing the adoption of new environmental and social measures. Cooperative efforts among different groups, a feature that enables societies to act more effectively, may or may not become commonplace. But these new and still tentative approaches represent a fascinating and promising phenomenon.

Moreover, the information revolution has undoubted potential to empower citizen's groups by allowing them to acquire, analyze, and dispense information and to communicate with other organizations. A few keystrokes can now disseminate information by e-mail to hundreds of different groups linked for a particular project; drafts of an agreement or a press release can be jointly prepared, edited, and approved within days and then simultaneously released all over the world. Indeed, my colleagues and I see how this technology has transformed our own work, since we communicate and share data with similar organizations in Africa, Europe, Latin America, and Asia virtually every day. So the social force represented by citizen's groups and community organizations may grow in power as the information revolution continues. Of course, hate groups can use the same tool, but their numbers are far smaller than those of environmental and social reform groups.

The information revolution could also conceivably transform the process of governance and the delivery of public services. A recent study by the Rand Corporation found that communities with universal access to e-mail had better communication within the community and between citizens and governments than communities that lack e-mail. The electronic links in turn led to better social integration; more access to employment opportunities, health services, and other useful information at very low cost; and more responsive governments.[10] Esther Dyson, a leading analyst of electronic networks such as the Internet and their uses, describes their potential social impact in this way: "People need the experience of building a society, of doing more than voting and paying taxes. . . . My great hope from the Internet is that it will involve a growing portion of the population in some kind of governance, and that their feeling of empowerment will then spread to other parts of their lives."[11]

One place where such a future can be glimpsed is Finland, which already has nearly universal access—it has more Internet hookups per person than anywhere on Earth, twice the U.S. number, and will have every school in the country on-line by the year 2000. The country's capital, Helsinki, is building an interactive network for the entire city, complete with cameras and audio links, allowing Finns to electronically drop in on concerts, university lecture halls, shops, government offices, and friends.[12] Electronic access is becoming an integral part of the democratic process in Finland, as much a guaranteed right as the right to vote.

What about Internet access in developing countries, especially those so poor as to make the Finnish situation seem like a dream? There, too, technology may offer a chance, for those countries who seize it, to provide more information, better services, and improved governance—in effect, to accelerate social development. Here, "wired" is probably not the right word—it may be a long time before Indian and African villages or even urban areas are well served even by conventional telephone lines—but cellular phones are becoming ubiquitous throughout Asia and parts of Latin America. They will eventually become inexpensive as well, since the market for them is huge.

Cellular phone technology will bring telecommunications to the urban middle class much sooner than would otherwise be possible; community-owned or cellular pay phones may do the same for poorer neighborhoods. Digital versions of cellular phones with small video screens are already able to carry fax and e-mail traffic as well, in effect making the Internet airborne. This is a powerful example of technological leapfrogging—short-circuiting earlier developmental patterns and taking societies into the cyberspace age before conventional telephones or even personal computers are in widespread use.

In 1996, Motorola launched the first of several novel satellite systems intended to extend cellular networks worldwide, which failed financially. But Teledesic LLC, backed by Microsoft's Bill Gates and cellular telephone magnate Craig McCaw, is readying a competing system, as is Globalstar. These will feature hundreds of satellites in orbits so close to Earth that special cellular phones and handheld computers will be able to connect directly to them—and hence communicate to or from anyplace on the surface of the planet. Potentially, then, as prices come down over the next decade or two, these systems could extend telephone links and e-mail networks even to isolated villages. The infrastructure will be there; less certain is whether agricultural, health, and other information useful to villagers will be accessible in appropriate languages and at low cost, although electronic repositories of such information are something that foundations and development agencies could easily provide.

Technology is only a tool, however powerful. Broader social trends, such as rising levels of education and the growing economic and social liberation of women, also raise hopes for a *Transformed World.* Although most models omit social factors from analytical studies of the future because they are difficult to quantify and thorny to predict, such factors are critical determinants of the future—and many of the current trends are in the right direction. A recent U.N. report asserts that if countries accelerate these trends with targeted programs, they could nearly eliminate severe poverty within twenty years.[13] Moreover, some social factors, such as public attitudes and political consensus, can change very rapidly—far more rapidly than population size or energy consumption. The peaceful transformation from white to black rule in South Africa—from a society ruled by apartheid to a society run by a broadly based democratic government—is one recent example; the rapid transformation of the Czech Republic, Poland, and Hungary to democratic governments and free-market economic systems is another. Even the turbulent, hesitant, but probably irreversible transformation of Russia from a closed, Communist dictatorship to a democratic form of government and a market economy within less than a decade shows how radically the social order can change for the better. Thus rapid social transformation cannot be overlooked as a potent possibility for the future.

To be sure, the *Transformed World* scenario is an optimistic vision of empowered citizenry, enlightened corporate actions, and radical policy changes. It depicts a society that relies on market forces but does not substitute them for deliberate social choices; that endorses competition but sees cooperation and solidarity among the world's peoples as equally important forces. It argues that social attitudes and norms can change dramatically so that societies can adopt priorities radically different from today's. Overall, *Transformed World* projects a global civilization that actively shapes its own future—in communities and nation by nation—to achieve the hope of a better life for all.

Such a future certainly will not come about without fundamental shifts in direction, without profound social and political change. But the *Transformed World* scenario and the regional versions given later in this book suggest that there is a chance, that there are abundant opportunities, to halt the drift toward a *Fortress World* and to create a more stable and equitable future, one shared by more of the world's people, than a *Market World* may be able to provide.

PART III

TRENDS THAT SHAPE OUR FUTURE

WHAT ARE THE CRITICAL LONG-TERM TRENDS that will shape the next half century? It clearly matters how many people inhabit Earth, how much and what they consume, and whether they are rich or poor—so demographic and economic trends are fundamental. Africa, for example, may face a tripling of its population in the next half century, whereas Russia's population could decline by nearly a third.

Another key determinant of the future is technological innovation. Radical shifts in technology can engender sweeping social and economic changes. And just as the harnessing of steam energy ushered in the industrial revolution, we are now living in the middle of a profound transformation brought about by the revolution in information technologies. The next fifty years will very likely also see major shifts in energy technologies and radically new possibilities created by biotechnology, high-tech materials, and other frontier areas.

Environmental changes could also prove critical, especially if they threaten human health or irreversibly alter the environment for future generations. Political and social transformations are important as well: the further spread of democracy, for example, or significant improvements in the status of women could make a big difference to the world of the future. And finally, it matters whether the world is at war or at peace, whether societies are secure or threatened by widespread violence and instability. All these factors deserve careful scrutiny to see where current trends are headed, toward what kind of world they point.

Trends are not destiny. Human societies have an enormous ability to shape their future and to adapt to changed circumstances. Over the past several decades, countries such as South Korea have risen rapidly from poverty and transformed themselves into modern industrial states while others, such as Nigeria, have squandered rich oil resources and sunk into a more wretched existence.[1] Such different trajectories underscore the power of choice, suggesting that the countries of the world are not locked in to any given future.

And yet in a certain sense trends *are* destiny in that they contain clues to—and constraints on—the shape of the future. Some trends have enormous momentum and will not be easily altered. Halting population growth, for example, could take a couple of generations as birthrates adjust to different economic and social conditions. Reversing the rising trend of greenhouse gas emissions might take many decades given the energy needs of rapidly industrializing countries and the huge investments in coal mines, oil refineries, and gas (or petrol) stations that help to lock in the world's addiction to oil and other fossil fuels.

Similarly, the poverty of hundreds of millions of people cannot be remedied overnight. Even if India, for example, were to experience very rapid economic growth, the country simply could not attain even half the present U.S. levels of prosperity within fifty years.[2] As rapidly industrializing economies expand their production, pollution is likely to rise, too, at least for a while. Production of chemicals in China, for example, may well increase tenfold in the next half century, and so will the risk of toxic emissions.

Systematically examining such trends—especially persistent trends[3]—can suggest the likely boundaries of a country's trajectory, its plausible range of development, and the scope of problems it may face over a given period. This technique, known as critical trends analysis, is a widely used tool for studying the future and is the basis for the chapters that follow.

Chapter 6

Critical Trends: Demographic, Economic, and Technological

EARTH IS NOW HOME to nearly 6 billion people, and the population is still growing rapidly. Since roughly 1 billion of those people are girls under the age of fifteen, most of whom will want children, it is inevitable that the human family will continue to expand for some decades, reaching 7 billion soon after the year 2010, 8 billion by 2025, and perhaps 9.5 billion by 2050. This is certainly a critical trend, one that will shape a more crowded future.

Global economic activity is nearing $30 trillion per year, growing at about 3 percent per year. If growth continues at this rate, economic activity will double at least twice in the next fifty years, reaching an annual volume of $120 trillion; average per capita income will more than double. Because such increases in prosperity could alter national destinies and improve people's lives enormously, economic growth also counts as a critical trend.

But what if the pattern of growth is uneven, so most of the new prosperity accrues to those who are already wealthy and gaps between rich and poor become yawning chasms? Or what if global industrial expansion, rising consumption, and swelling populations also raise pollution levels worldwide and accelerate other forms of environmental degradation? These trends, too, could profoundly alter, perhaps irreversibly, the world that future generations will inhabit and must also be considered critical. Threats to human security—not only age-old scourges such as war and hunger but also emerging concerns such as terrorism, global crime, and new diseases—may also prove critical.

Set against such grim portents are more optimistic social and political trends such as growing levels of education, the slow but seemingly inevitable rise in the status of women around the world, and the spread of democracy. New technologies—medical breakthroughs, worldwide communications systems, radically more efficient cars, less-polluting industrial processes—loom on the horizon. These also will play a critical role in shaping the future.

Many of the critical trends considered here describe a world in which conditions are getting worse, not better. I wish it were not true, but that is what the data all too clearly show. Environmental conditions, for example, may be improving in the United States and in a few other wealthy countries, but globally they are deteriorating rapidly. Although many U.S. citizens are enjoying unprecedented prosperity, the number of people in the world living in destitution and poverty is rising. Moreover, the analysis underlying this book suggests that if current trends continue, many of these conditions will get far worse. Within a decade, for example, more than half the world's children will grow up in the teeming, congested, mushrooming cities of the developing world—for many, squalid, polluted slums will be their only playground. Within two decades, shortages of water—especially water clean enough to drink and to cook with—will have reached crisis proportions in several dozen countries and irreversible loss of coral reefs and tropical forests will have attained alarming levels. Within three decades, unless thousands of climate scientists are badly wrong, evidence of a changing climate caused by our industrial civilization will be unmistakably provided by unprecedented floods, droughts, and heat waves.

To repeat myself, trends are not destiny, and they should not be mistaken for predictions. Not only can the truly unexpected happen, but also, ultimately, the underlying forces that drive these trends are human processes and depend on human decisions yet to be made. Nonetheless, much of the section that follows will be depressing to read. But unless we understand the potential for both good and bad, planners, government officials, and citizens may not fully comprehend what needs to be done and thus may not take action to reverse negative trends and bolster positive ones. So it is important to analyze these trends, to understand their causes and implications, and to explore plausible trajectories into the future if we are to help shape our destinies and choose a more hopeful future.

What sets these critical trends apart from more ephemeral trends in human affairs—styles of dress, the popularity of particular sports, the ups and downs of the business cycle—is not just their importance but also their staying power. The critical trends described here reflect powerful underlying forces that drive fundamental change; hence, the trends are not easily altered and tend to persist over long periods of time. Think of the several million scientists and engineers worldwide creating new knowledge, new materials, and new techniques and it is easy to see why new technologies continue to appear. A charismatic public figure can almost single-handedly alter popular pastimes—as demonstrated by Michael Jordan's impact on basketball, a sport now enjoying unprecedented worldwide popularity—but no one person and no political edict can slow population growth overnight.

Persistence is important for our purposes because it means that many of these critical trends can be projected into the future with some confidence, either directly, by extending a line on a graph, or in a more sophisticated way, with a model that takes into account a number of variables. Given the uncertainties of any projection, especially over a fifty-year period, it makes sense to consider a range of projections that together map out a set of plausible conditions for the future. The United Nations, for example, routinely issues high, medium, and low projections for the populations of most countries, and I will follow the same approach. Even with some uncertainties, such projections provide a powerful tool for thinking about the future.

Analysis of critical trends provides quantitative guidelines or constraints to go along with more qualitative insights from scenarios; projected trends provide a basis for comparing the prospects of different regions and for separating plausible from implausible futures. They can help us decide where we are headed—toward a *Market World* future or a *Fortress World*—or what may be required for a *Transformed World* to be plausible.

Population Growth and Decline

Over the past half century, the world's population has grown by about 2 percent per year, doubling in just thirty-seven years and increasing from about 2.5 billion in 1950 to 6 billion by the year 2000. But in nearly every country, women are choosing to have fewer children than their mothers did, so families are smaller. This decline in what demographers call total fertility, the average number of children per woman, means that the world's population growth rate is slowing. Seen in historical perspective, however, the number of people who live on Earth is still rising rapidly (see figure 1), we are still on the steepest part of the population growth curve. The reason is sheer numbers: every day, there are more people to have children.

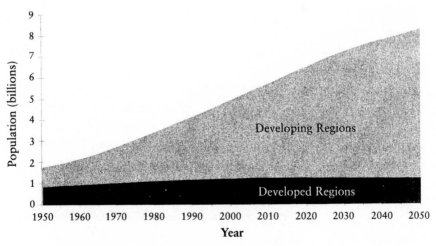

Figure 1. Projected Growth in World Population

Most developing countries still have relatively high fertility and consequently are in the middle of a baby boom. These are societies dominated demographically by young people: large numbers of women still in their childbearing years or under the age of fifteen. This amounts to a built-in guarantee that population growth will continue for some time to come, even as women individually choose to have fewer and fewer children. Thus, the world's population is expected to swell to 9.4 billion by the year 2050 (the U.N. medium growth projection), even allowing for declining birthrates and the AIDS epidemic. And nearly all those additional people will be born in developing regions of the world.

Eventually, for Earth's population to stabilize, women around the world must produce no more than 2 children each on average—in effect, each couple replaces itself. This is what demographers call a population's "replacement rate" fertility.[1] Industrial societies have already lowered their fertility that far or further: women in the United States are reproducing below the replacement rate, and women in western Europe and Japan average only 1.5 children each.

In developing regions, the pattern is more varied. China has already reached replacement rate fertility—a consequence of that nation's tough birth control policy, which penalizes couples who have more than one child. Even so, its population will not fully stabilize for another two generations. But in much of Africa, women still average close to six children each. Other regions are somewhere in between (see figure 2).

The U.N. medium growth projection shown in figure 1 assumes that fertility will decline to the replacement rate in every country by or before the middle of the next century. That is a big assumption. It includes the implicit assumption that every country successfully develops—that poverty declines, that education levels (especially for women) rise, and that women everywhere choose (and are allowed to choose) to have only two children.

Even slightly higher fertility rates have enormous long-term consequences. If fertility stabilized at a global average of 2.5 children per woman instead of 2.1, the world's population would exceed 11 billion in the year 2050 (the U.N. high growth projection). If fertility declined more rapidly than expected and reached 1.7 children per woman, world

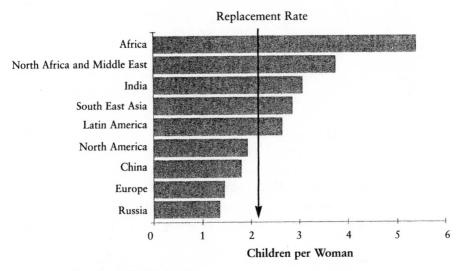

Figure 2. Total Fertility by Region

population would peak around 2040 and decline slightly to about 7.6 billion by midcentury (the U.N. low growth projection). The difference is huge—3.5 billion people, as many as now live in all of Asia. So it matters a great deal how soon fertility declines and whether it reaches the replacement rate, which will allow populations to stabilize.

The high and low projections mark out the *plausible range* of population for the year 2050, barring truly unexpected events. But global numbers tend, if anything, to mask the consequences of continued high fertility or low fertility in particular regions. It would be fair to say, in fact, that there is no global population problem. Instead, there are a number of acute—and different—regional population problems.

The sub-Saharan region of Africa today, for example, is home to just fewer than 600 million people. If Africa's fertility rate were to decline more gradually than expected and level off at 2.5 children per woman, that region's population would grow to a staggering 2.1 billion people by the year 2050; if it were to follow the low growth projection, it would reach a still large 1.5 billion. So although the number of people living in the region is expected to roughly triple over the next half century, the plausible range is anywhere from 2.5 to 3.5 times the present population. Such numbers pose enormous problems for a region that even now cannot feed its people adequately.

In contrast, China's population a half century hence could be slightly smaller than at present, about 1.2 billion, or it could swell to more than 1.75 billion. India is likely to surpass China, becoming the world's most populous country, but its plausible range is from 1.2 billion and 1.9 billion. Experts worry, however, that if either China's or India's population reaches the high end of the range, agricultural production could fail to keep pace and either country could become such a massive importer of food as to drive up world prices dramatically and perhaps even cause global shortages. In the volatile region of North Africa and the Middle East, population could nearly triple, putting an enormous additional strain on a region already short of water and beset by other tensions. (The appendix gives detailed population projections—high, medium, and low—for each region.)

Of course, conditions may change. The number of people Africa could support if it were largely urbanized, with a strong industrial base and modern agriculture, would be far larger than could be supported by today's Africa, which is largely rural and has vast numbers of people still dependent on subsistence agriculture. The same is true for India and China. So the interesting question is how rapidly these regions will develop and in what way. Nonetheless, how fast populations grow and how soon they stabilize will powerfully shape the future of these regions and, inevitably, of the whole world.

Demographically, the industrial regions and the transitional regions of eastern Europe and Russia face just the opposite trend. Populations in Europe and Japan are beginning what appears to be a slow decline. Populations are declining even more dramatically in eastern Europe and, especially, in Russia. Only in the United States is the population still growing significantly, in large part because of continuing immigration.

What implications do these contrasting demographic trends have for the future? To begin with, populations throughout the industrial countries are rapidly aging; the "old old," those eighty and older, are the most rapidly growing segment of the population. That means larger numbers of retirees, with higher medical costs and fewer workers to support them. These demographic changes are likely to break Europe's generous social contract and require countries in the region to prune back benefits they can no longer afford, a prospect that is already triggering strikes and political controversy in France. The United States faces similar long-term trends—witness Americans' growing concern that the social security system will collapse when the baby boom generation starts to retire—but the trends are moderated to some degree by higher birthrates and continuing immigration. The aging of these regions may have social, political, and economic consequences too. Will aging societies be more conservative and less dynamic than the youth-oriented America of the past fifty years? Will social and technological innovation shift to the more youthful cultures of the developing world so that the world of 2050 will perceive that the future happens first in Santiago, Chile, or Bangalore, India, rather than California?[2]

The sheer weight of numbers will also have an effect on the political and cultural makeup of the future. The population of the industrial countries will shrink dramatically relative to those of the rest of the world. By the year 2050, for example, the United States and Canada will be home to less than 5 percent of the world's population; the world will be overwhelmingly Asian, African, and Latin. Will English still be the dominant language of science and international business, as it is now, or will Chinese, Spanish, and other languages have moved to the fore? And even if English prevails, will other cultural artifacts, such as movies and popular music, eventually follow the tastes of the potentially far larger audiences (especially the youthful audiences) in the emerging parts of the world?

Where people live is also important, and that is changing rapidly as the developing world undergoes an urban explosion. Twenty-five years ago, two-thirds of the world lived in rural areas. Not long after the start of the twenty-first century, half of the world's population will be urban; in another twenty-five years, two-thirds will be. Indeed, over the next fifty years, more than 90 percent of the additional people expected on Earth will end up in cities in developing countries—there probably will be at least 3 billion new urban dwellers.

This massive surge of urbanization will transform the character of developing countries such as China and India, both now still 70 percent rural. Worldwide, urban migrants already account for a million new city dwellers every week—a rate of growth that far exceeds the pace at which proper housing can be built and sewage systems installed. For example, as many as 100 million people in China are thought to be on the move from rural to urban areas—a transient population that is already crowding train stations and urban shantytowns to intolerable levels. Slum cities inhabited by urban migrants surround virtually every major city in Africa, Latin America, and Asia. And cities are reaching enormous sizes—27 million people in Tokyo, 16 million in São Paulo, 15 million in Bombay—making them hard to manage and placing enormous strain on their local environments. And yet the urban surge is still in its early stages and is expected to accelerate.

One way to think about this trend in human terms is to imagine what it means for most of the world's children to grow up in urban areas, many of them in the crowded cities of developing countries. The urban context will shape, for better or for worse, the sensibilities and perceptions of most of the people born in the twenty-first century.

———

Regardless of where people live, can the world support a population 50 to 80 percent larger than it is today? Can we feed and house and transport—to say nothing of providing consumer luxuries for—that many additional billions? Most agricultural scientists believe it is possible to grow twice as much food as is grown at present.[3] What is harder, and less likely, is to get food to all who need it. The number of chronically hungry people in sub-Saharan Africa, for example, is projected to nearly double—to 300 million—within just fifteen years. If the current trends continue, one out of every four children worldwide will be malnourished in the year 2020.[4]

But there is another way to answer the question. If people could afford to buy the food they need, the experts say, then it is likely that the world could grow it, albeit at some environmental cost. So the question is partly an economic and social one—whether poverty can be done away with, whether incomes will rise; in short, whether development will succeed. In part, that will depend on the rate of economic growth, the second of the critical trends that will shape the future.

Economic Growth

Over the past fifty years, global economic activity has outpaced population growth, although not in every country. The world's economy has grown by nearly 3 percent per year, doubling twice in that period. If this trend continues for the next fifty years, economic activity will double twice more. Some analysts think that the world economy might conceivably expand sixfold over the next half century (an average growth rate of about 3.5 percent).[5] Since economic activity is now growing more than twice as fast as population, Earth's inhabitants in the year 2050 will be richer—or at least they would be if all people and all regions shared in that economic growth.

China, with its almost miraculous transformation from a centralized economy to a partial market economy, illustrates the power of economic growth to change a nation's destiny and improve the lives of its citizens. The economy has grown by nearly 10 percent per year for the past fifteen years, despite dragging the deadweight of its state-controlled heavy industry sector; the Shanghai area and the southern coastal provinces near Hong Kong have grown by nearly 20 percent annually. As a result, China now has a small clutch of millionaires, a growing urban middle class, and a sense of growing prosperity in many parts of the country; incomes for most Chinese have tripled in fifteen years. Nor is China alone. Economic growth has surged in Thailand, Indonesia, Malaysia, and Chile, to name some of the past decade's brightest stars. South Korea, Taiwan, and Singapore—poor countries just a few decades ago—are now established industrial powers with near European income, a remarkable achievement despite the 1997–1998 financial crisis affecting South Korea and many other Southeast Asian countries. Once the crisis passes, growth in these countries is likely to resume. Even with the current economic crisis, growing pollution problems, and sometimes repressive governments, it would appear that the quality of life is clearly improving for many of the citizens of these countries. Why have many African countries and some in Central America and the Middle East grown so slowly or sporadically that—in more than 70 countries—average per capita incomes have actually declined since 1980?

Economists still argue about the causes of economic growth. Historically, trade has often led to prosperity. So have manufacturing improvements. But in today's world, new technology and the high-tech skills of highly educated knowledge workers appear to be far more important in the creation of new products and services—and of prosperity—than traditional inputs such as natural resources, labor, and capital. Indeed, many cite the lack of an educated workforce and lack of effective legal and government institutions as among the primary reasons why economic growth has remained elusive in some developing countries. Regardless of the reason, such countries demonstrate that economic growth is not inevitable. The recent stumbles of many fast-growing Southeast Asian countries and the U.S. surge of inflation in the late 1970s illustrate that flawed policies can humble even the most robust economy.

But just suppose the magic of economic growth spreads to more countries, allowing a *Market World* scenario to unfold rapidly in nearly every region. The World Bank recently forecast that developing regions would grow by more than 5 percent per year over the next decade, twice the average rate of growth of the industrial regions.[6] If growth at nearly that rate continued for half a century, what would the world look like in 2050?[7] The result would be more than a sevenfold increase in the global economy and a significant shift in economic power, with the economic output of Europe, North America, and Japan more than matched by that of Asia, Latin America, and Africa (see figure 3). These are deliberately optimistic projections that establish an upper bound to the economic trajectory of each region over the next half century. A far more dismal picture emerges if economic growth is slow or sporadic, averaging only 2 percent annually in developing regions. Such minimal growth projections—only just keeping up with population growth in sub-Saharan Africa, for example, thus obliterating any hope of improved prosperity for the region—provide a lower bound to each region's prospects. Midrange economic projections lie somewhere in between these extremes. Those used in this book assume a continuation of recent economic patterns in most regions and add up to slightly more than a fourfold expansion of worldwide economic activity by the year 2050. (Detailed economic projections for each region and the assumptions on which they are based are given in the appendix.)

Just as with population, the high and low economic projections set

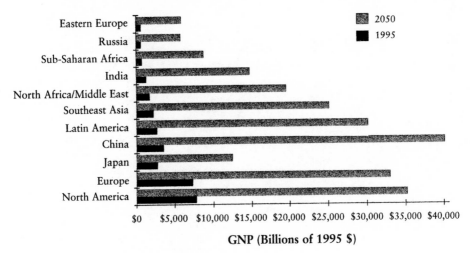

Figure 3. Projected Economic Growth

out a plausible range for economic activity in a given region. Combined with population projections, they indicate how rapidly average per capita gross national product (GNP)—the standard measure of the state of economic development and, for our purposes, a rough measure of average per capita income—might rise. The best case for developing regions would be one of high economic growth and low population growth, whereas the worst case would combine low economic growth and high population growth. The range between these extremes is extremely wide: half a century from now, average incomes in Latin America might conceivably reach $46,000 per person or might be held back to as little as $8,000 per person. Similarly wide ranges are projected for Southeast Asia and China (see table 1). Clearly, it makes a huge difference to the long-term prosperity of a region which economic and demographic trajectories it follows. But that, as we will see, in turn depends on environmental constraints and on social and political factors—that is, it depends on which scenario a region follows.

The plausible range itself, more than any specific projection, gives a strong indication of how rapidly economic development might proceed in one region as compared with another. In India, for example, the plausible range of average incomes half a century hence is $2,100 to

Table 1. *Plausible Range of Average Annual Per Capita Income in 2050*

	WORST CASE	BEST CASE
Developing Regions		
Latin America	$8,000	$46,900
China	$6,000	$33,500
Southeast Asia	$6,600	$37,800
India	$2,100	$11,900
Sub-Saharan Africa	$1,100	$5,700
North Africa and Middle East	$5,500	$30,000
Transitional Regions		
Russia	$20,400	$40,500
Eastern Europe	$18,200	$37,600

Note: See Appendix for assumptions and economic projections on which these figures are based.

$11,900 per person; in sub-Saharan Africa, $1,100 and perhaps $5,700 per person.[8] Over the next half century, sub-Saharan Africa can at best aspire to the level of development, the average prosperity, of Latin America today and must struggle just to raise the majority of its people above the poverty line. These are harsh constraints.

The struggle to stimulate and maintain economic growth in one country or another might seem like challenge enough. But in today's increasingly mobile and interconnected world, the patterns of economic growth within a country and among regions can also influence the course of events and the movement of people. And here, the trend toward increasing disparity in wealth and income—our third critical trend—is a worrisome one.

Issues of Equity

Waiting lists for legal migration into the United States from Mexico, the Philippines, and China are filled for years to come. Every week, U.S. border guards catch hundreds, sometimes thousands, of people entering the country illegally. Some of these would-be migrants die before reaching the border, swept away in the treacherous currents of the Rio

Part One

How these papers have been placed in sequence will be made manifest in the reading of them. All needless matters have been eliminated, so that a history almost at variance with the possibilities of later-day belief may stand forth as simple fact. There is throughout no statement of past things wherein memory may err, for all the records chosen are exactly contemporary, given from the stand-points and within the range of knowledge of those who made them.

— Bram Stoker, *Dracula*, 1897

Grande, which divides Texas and Mexico, or perishing in the crowded hold of a tramp steamer en route from southern China. But nearly half a million per year succeed, according to official estimates, and the number may well be higher.

Western Europe, too, has been flooded by migrants from the countries of the former Soviet Union and from Africa. Tens of thousands of Chinese seek to enter prosperous Hong Kong every year despite determined attempts to keep them out—first by British authorities and now by the Chinese government itself. The eastern states of India—poor and crowded though they may appear—nonetheless have attracted millions of still poorer, often landless migrants from Bangladesh. Post-apartheid South Africa has become a beacon for much of that continent, attracting huge numbers of unwanted settlers.

The motivation for most of this desperate human tide is simple: economic opportunity. Not only are there jobs to be had, but even the dirty, hard jobs available to most new illegal migrants in the United States pay far more than the average income available in, say, an El Salvadoran village. A woman working for a few years as a prostitute in Italy—where most streetwalkers now come from Albania or Nigeria—can earn enough to buy a house or establish a small business back home.

Global disparities in wealth and income are rising rapidly, widening the gap between rich regions and poor ones. Between 1970 and 1990, for example, the gap between average per capita income in developing countries and that in industrial countries doubled from less than $9,000 per person to nearly $18,000 per person (measured in constant dollars).[9] Already, a U.S. resident with an annual income of $35,000 earns more in a year or two than a poor inhabitant of rural India or Africa will in a lifetime, even when those incomes are adjusted for their real purchasing power. Is it any wonder that many young people in the developing world leave their homeland, bound for a more prosperous region? But these gaps in income are likely to become even more pronounced in coming decades, increasingly dividing human societies into two: one of wealth, luxury, and power; the other of poverty, hardship, and often hopelessness.

To illustrate these emerging divisions, consider what happens using the most optimistic projection of economic and population growth in developing regions along with medium projections for industrial regions—not the most extreme case by a long way. By the year 2050, according to these projections, average per capita incomes in the United States and other industrial countries will reach about $64,000 compared with less than $5,700 in Africa and $11,900 in India. Indeed, the projections show that the gap between average incomes of people in rich countries and those in developing countries, measured in dollars, will steadily widen over the next half century (see figure 4).[10]

These trends can be looked at in another way. According to the United Nations Development Programme, the income of the richest fifth of the world's population was thirty-two times that of the poorest fifth in 1970; by 1990, the top fifth was earning sixty times as much as the bottom fifth.[11] By 2025, if current trends continue, the disparity between the top and bottom fifths will be more than 100-fold; by 2050, depending on the projections used, nearly 200-fold.

Gaps between rich and poor are also increasing *within* many countries. Indeed, they are often most starkly on display in the urban centers

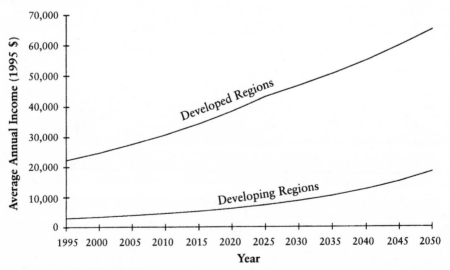

Figure 4. Income Gaps

of developing countries, where shantytowns sometimes abut luxury hotels and apartments. Latin America has the highest income disparities of any region: in Brazil, the top fifth of the population has thirty-two times the income of the bottom fifth.[12] Incomes in China and Russia, largely equitable societies in the Communist era, are now rapidly becoming more unequal: urban incomes in China are now three times rural incomes and rising twice as fast, which in part accounts for the flood of urban migrants in that country.[13]

But rising gaps between rich and poor are not restricted to developing countries. The United States, despite its egalitarian rhetoric, has become increasingly inequitable in the past two decades—it is the most unequal of all industrial societies in concentration of wealth, in distribution of income, and in access to health care and other social amenities.[14] Those on the bottom, officially counted as living below the poverty line, now account for nearly a fifth of all U.S. households; this growing form of economic segregation is hard to reconcile with the image of the United States as a just and fair society.

Might disparities within countries—in China, in Brazil, in the United States, or elsewhere—pose a growing threat to social stability and to any country's long-term economic fortunes? So far, the evidence is ambiguous, although some economists argue that economic growth is faster in more equitable countries. Certainly, raising a significant fraction of U.S. children—that is, of the future U.S. workforce—in poverty does not sound like a recipe for social harmony or for successful competition in the global economy.

But the already large and prospectively enormous gaps in income *between* countries clearly will have an effect on the aspirations and the movement of people. The differences in lifestyles and opportunities available to the rich and those available to the poor are increasingly visible through global media and widespread tourism, leading, as already described, to pressure for migration. Such pressures will almost certainly accelerate and may threaten political and social stability in recipient countries—there are already early signs of a political backlash, for example, in France and in parts of the United States. Enormous disparities can also lead to despair and hopelessness or can help fuel anger and a sense of injustice that may add to emerging new security threats such as terrorism or crime targeting tourists, such as the 1997 slaughter of tourists in Egypt.

======

Could advanced electronic sensors help to slow illegal migration or detect and defeat terrorists? Could miracle technologies not yet invented—new contraceptives, perhaps, or novel ways to grow food or new sources of energy—slow population growth more quickly or solve some of the other problems implied by the critical trends under scrutiny here? Indeed, in this high-tech age, the belief that technological innovation—our next critical trend—will transform the future is widespread and understandable. But is it correct?

Technology Trends

Rapid technological change has characterized the past half century. During that period, the number of scientists and engineers in the world—human agents of innovation—has grown by about 6 percent per year, three times as fast as Earth's population, increasing by more than a factor of fifteen. Even if that trend slows somewhat, the rapid growth of knowledge and innovation is likely to continue. Our grandchildren and great-grandchildren will almost certainly take for granted new technologies (and improvements in familiar technologies) that we today might find almost magical, just as people in the post–World War II era would have been astonished by pocket-sized cellular telephones, laptop computers, and cloned sheep.

But even though such changes will indeed transform the world, they may not necessarily solve fundamental demographic, social, and environmental problems—just as the advent of computers, cellular phones, and even television did not do so over the past fifty years. The effect of technology, after all, depends on its social context—it is how, and whether, we use it that counts. Can we say anything, then, about the effect of another fifty years of rapid technological change?

At the very least, technological innovation will have important economic effects—lowering costs through improved efficiency and accelerating economic growth—as discussed in the *Market World* chapter.

Even so, a recent study concludes that inventing new technology is less important than using it effectively.[15] This also means that regions of the world without access to this continuing flow of knowledge, or lacking the social organization to use it well, will fall further behind.

Advances in contraception and in health care in general—certainly plausible in the next half century—would help couples who want only two children to limit their family size as desired, but lack of access to contraception is not the most important factor behind large families even now. Poverty, educational levels, the status of women, and cultural factors such as a preference for large families are far more important.

Might technological breakthroughs solve some environmental problems? Possibly, but here social factors play a role, since it usually takes decades or longer for new technology to be adopted and to begin to have its full effect. Imagine that a pollution-free car costing no more than a traditional model were available now. Absent regulations requiring its use, the car might well take at least a decade to become popular and trusted enough to capture a significant share of the new-car market, and then it could take at least another ten years, given the slow turnover of the automotive fleet, to account for a majority of the cars on the road. So the full effect of such a breakthrough might not be felt before 2025, even in the industrial countries. Power plants that generate electricity last forty years, not just ten or fifteen, so breakthroughs in energy production could take even longer to have a significant effect—at least in a *Market World* scenario.[16] In the meantime, the number of cars and power plants are growing rapidly to meet the expanding needs of developing economies and, in their sheer number, may simply overwhelm the effect of cleaner technologies so that global pollution levels continue to climb.

These time factors—what we might think of as a societal "learning curve" for technology—mean that most of the critical new technologies that might have a major influence in the next twenty-five years and perhaps the next fifty years are already here, in laboratories or in limited use. These certainly include the information technologies and biotechnologies and probably include new energy technologies such as solar cells, fuel cells, and new high-tech materials designed almost literally molecule by molecule for particular tasks. The social and economic effects of these technologies may hold some surprises, since most of these effects—even for the information technologies—are still to come. And some truly radical innovations may emerge, although they too are likely to be subject to the societal learning curve, meaning that they will really matter only toward the end of the next half century. But to end poverty or human conflict or to turn around other critical trends will almost certainly take social as well as technological innovations.

Some observers worry that rapid shifts in technology, far from solving critical social problems, might actually make them worse. The information revolution, for example, is certainly contributing to a rapid shift away from low-skilled jobs in favor of more knowledge-intensive jobs such as those in the fields of health, finance, and software. Some analysts argue that these changes and the increasing ability of global corporations to shift work wherever wages are lowest are undermining the job market throughout the industrial world.[17] Others worry that information technologies and the industrial restructuring they have made possible are eliminating many jobs, not just shifting their nature.[18] Historically, technological change has usually increased, not decreased, employment, as measured over decades. Will the continuing information revolution prove otherwise? To date, the evidence suggests not: there has been relatively little unemployment in the United States, where industrial restructuring has gone furthest. And in Europe, where restructuring is less advanced and labor markets more rigid, unemployment levels are high. But there is no denying that rapid shifts in labor markets can be painful to the individuals affected.

The modernization of agriculture—itself a technological revolution for the farmers involved—poses a similar threat in developing regions. Industrialization will create many new jobs, but whether there will be enough to absorb the huge numbers of surplus rural workers caused by agricultural modernization remains uncertain. And large-scale unemployment, especially in urban areas, could lead to social unrest and upheaval, a particular concern in countries such as China.

New technologies do offer hope that they will also kindle novel, and favorable, social developments. Computers and inexpensive communications, visionaries point out, could potentially connect all of humanity—becoming, in effect, the nervous system for a global civilization.

Such a development, unparalleled in human history, could speed development, bind communities and nations together, and enrich lives. Or it could instead undermine traditional values and cultures, fragment communities, and create new disparities in the form of information haves and have-nots. Whether human societies can successfully use these new tools—and other new technologies and forms of human ingenuity—to craft solutions to emerging social and environmental problems is far from certain. But the potential is there. On balance, the onrush of new technology is a positive critical trend.

Interactions

The critical trends examined here and in later chapters do not operate independently of one another. What happens with respect to one issue often reverberates in other trends—the world is in fact a complex, interconnected, adaptive system in the sense described by Murray Gell-Mann and other theorists of the science of complexity. Global population, for example, probably will not stabilize in the coming half century unless poverty is also conquered because poor couples have a strong economic incentive to have large numbers of children. But eliminating poverty in turn requires economic growth in poor regions and a fairer distribution of its benefits so that nearly all people can meet their basic needs. Such a scenario won't automatically unfold, even in a *Market World*. Capitalism, after all, is not about fairness; as Wall Street pundits remind us, it is about making a profit. But even in a more equitable world, poverty will persist unless Earth's biological resources are protected, since much of the world's population—especially in poorer regions—still depends directly on forests, fisheries, and soils for a livelihood. Yet preserving these resources while still depending on them to meet rising human needs will not be easy.

Or take the industrial expansion that economic growth will entail. If current patterns of consumption prevail, more cars and other accoutrements of modern lifestyles, including the resources to produce them and the energy to operate them, will be in demand. A world consuming two or three times as much energy and material goods *could* mean two or three times as much industrial waste and pollution—unless effective governments and judicial institutions enforce environmental regulations and promote cleaner technologies. Such institutions are weak in many parts of the developing world, yet the alternative to the social and political development that nurtures them may be a far more polluted world, perhaps even an irrevocably degraded environment to pass on to future generations.

In short, there is growing recognition that environmental, social, and economic concerns are closely linked and that economic development pursued in isolation might well fail as it has, so far, in most of the poor countries of the world. To achieve an optimistic future, a *Transformed World*, human societies may have to make some deliberate and potentially wrenching changes in current trends. The alternative may be to risk a *Fortress World,* and in such a future, as the scenario suggests, it will not be just the poor who suffer. Both rich and poor share the same planet, so global environmental degradation will affect both to some degree; and the misery of impoverished and unstable regions may spill over into prosperous regions. A Middle East again in flames, whether from conflicts over oil or (more likely) water or from internal conflicts, would not only bring local hardship and misery but would also disrupt a world economy that is likely to be even more dependent on the region's oil than in the past. An unstable China split by regional tensions or in conflict with its neighbors would also rock global markets and place military forces to alert. Human destinies, in other words, are intertwined.

Chapter 7

Critical Environmental Trends

AN EXPANDED GLOBAL ECONOMY will mean expanded industrial production. If current patterns continue, that will also mean large increases in wastes and pollution worldwide. Might this cause irreversible changes in climate or other forms of environmental degradation that could undercut economic and social progress? A more crowded world, especially in the poorest regions where populations are still overwhelmingly rural, will need more of the basics of life—water, food, cooking fuel, shelter. Might these needs overwhelm the available resources—the supplies of fertile land, pastures for grazing, and firewood—that sustain villages and nomadic communities? Might these trends together put such stresses on natural ecosystems that some begin to fail, no longer able to support many desirable plant and animal species or to provide other essential ecosystem services?[1] The answers to such questions will help shape human destinies—help determine whether optimistic or pessimistic scenarios unfold, especially in

developing regions, as well as whether future generations everywhere
will inherit a bountiful natural world or one that is biologically impoverished.

Environmental trends are thus critical trends. But they are closely
intertwined with economic, demographic, and social trends and—
absent a sudden worldwide environmental disaster—may have their
most profound impact on the future trajectory of human societies in
indirect ways, as we shall see. And because such indirect effects are the
principal focus of this chapter, I do not discuss many of the most visible and most lamented changes to the natural world that may occur
in the coming half century: the virtual extinction of tigers and other
endangered large predators in the wild as their natural habitats
are appropriated for human use; the destruction of the large East
African herds of antelopes and wildebeests, which may go the way of
the North American bison as African grasslands give way to farmland;
the severe decline of migratory songbirds in many regions; and the loss
to pollution or development of many special, awe-inspiring natural
sites, from uncut verdant forests to vibrant coral reefs. These losses, if
they occur as now seems likely, are symptoms of biological impoverishment, and they will make the world aesthetically poorer, but they
will not in themselves alter human destinies. What might do so, however, are the two broad clusters of trends hinted at earlier: global industrialization and the environmental changes associated with it and
impoverishment of both rural communities and the environmental
resources they depend on.

These critical trends are largely invisible to most residents of industrial countries. Over the past few decades, for example, local environmental conditions have improved in the United States, Europe, and
Japan—air and water pollution are down, waste is under better control,
and there is a strong political consensus in support of these achievements. But in other parts of the world that lack such consensus or even
well-established and well-enforced environmental laws, environmental
conditions are worsening. Nearly all developing countries are focused
far more on growth than on environmental concerns. As a result, pollution is ubiquitous and often horrendous, sometimes comparable to
conditions in the industrial regions more than a century ago, an era

when raw sewage ran in the streets, when there were killer smogs in London and Pittsburgh, when industrial workplaces were often extremely hazardous. In 1997, for example, most of Southeast Asia was shrouded for weeks in a poisonous smog so dense that in some cities visibility was reduced to a few feet. Schools, offices, and airports were closed, and with the air pollution index reaching 839 in one Malaysian city (500 is the level deemed seriously harmful to health), the World Health Organization warned of respiratory and heart damage.

Yet industrialization is accelerating. If present trends continue, pollution could severely blight developing regions and threaten the health of large numbers of people, as it already has in many parts of eastern Europe and Russia.

But local pollution is only part of the problem. The continued buildup of greenhouse gases in the atmosphere, invisible to the naked eye, nonetheless has the potential to significantly change Earth's climate. And here, the industrial countries are overwhelmingly the major contributors, although emissions from developing countries are rising rapidly.[2] Combustion of coal, oil, and natural gas, the primary source of greenhouse gas emissions, is rising steadily as industrialization and urban lifestyles spread. If current trends continue, temperatures, rainfall patterns, variability and intensity of weather phenomena, and sea levels could change, not just temporarily but for centuries to come. The effects of such changes on human destinies will vary by region, but they could be profound—even forcing the permanent evacuation of some island nations and low-lying continental areas or the abandonment of agriculture in drought-stricken districts. Just a one-meter rise in sea level, plausible by the end of the twenty-first century, would drown as much as 80 percent of the Marshall Islands, displace 70 million people in Bangladesh, and inundate coastal areas in China, Japan, and the southeastern United States.

Impoverishment of rural communities and the environmental resources they depend on is also invisible, not just to inhabitants of rich countries but also to middle-class urban dwellers nearly everywhere because they don't depend directly on environmental resources for their livelihood or daily existence. But most people in rural areas of developing regions are dependent on what they can grow, gather, or catch.

The continuing degradation or destruction of aquifers, forests, soils, fisheries, and other environmental resources—by the rural communities themselves or by others—thus has a direct, negative effect on their welfare. These trends could further impoverish rural communities and impair development and could cause large-scale migration or contribute to instability.

Is rising pollution an inevitable consequence of a *Market World* future? Might a growing loss of renewable resources and increasing rural impoverishment help tip some regions into a *Fortress World* future? Just how might these critical environmental trends constrain the future of particular regions?

Industrial Expansion

Several years ago, the *New York Times* carried a report about a family in Bangkok, the booming capital of Thailand.[3] Richard Frankel had planned to leave Bangkok the evening before the start of a four-day national holiday, hoping to beat the traffic out of town. With the car packed and the kids asleep in the back seat, he and his wife left their house about ten o'clock in the evening. When they hit the expressway, however, they found themselves in a horrendous traffic jam that stretched bumper to bumper for sixty miles. By ten o'clock the next morning, they had only reached the airport, just a few miles from their house. They eventually managed to reach an exit and returned home to spend the holiday there. Some drivers abandoned their cars.

Unfortunately, traffic jams are not uncommon in Bangkok, and a three-hour commute is regarded as normal. The city boasts more Mercedes-Benz automobiles per capita than anywhere else in the world, but these days many of their owners are having portable potties and mobile phones installed in their cars, since gridlock often strikes without warning.[4] And it is not only congestion that makes Bangkok a model for an industrial civilization run amok. There are few public services—no subway, no garbage collection, very limited sewage treatment—and there are no effective environmental controls on motor vehicles, waste disposal, or air pollution. Most of Bangkok's inhabitants throw their garbage and flush their toilets directly into the city's canals. Booming industrial factories also freely discharge their waste into the air and the waterways. A 1992 study found that Bangkok's air exceeded World Health Organization guidelines for lead pollution by more than a factor of two and exceeded guidelines for suspended particulate matter by more than a factor of three.[5] Corruption is rampant, which means that businesses can readily and cheaply buy their way out of environmental or zoning regulations. This is the dark side of the Asian economic miracle, replicated in virtually every newly industrializing region.

As mentioned earlier, just how extensive and how damaging such pollution can be was dramatically illustrated in the fall of 1997, when a thick cloud of smoke and smog descended over a 5.18 million square-kilometer (2 million square mile) region of Southeast Asia. Tens of thousands of people were treated for respiratory illness, schools were closed, and tourists and foreign businesspeople departed. The smog, which lingered for several months, was caused by massive illegal burning of forestland in Indonesia, a common practice of logging firms and companies seeking to clear land for other purposes.

Rising pollution is not just a matter of unregulated capitalism, however. Expanding industrial activity requires energy, metals, chemicals, and other raw materials. Even the best of current technologies cannot completely eliminate pollution and wastes because they are, for the most part, predicated on a "once-through" materials cycle—from extraction to use to disposal.

Consider the most advanced model of a global industrial society available today—namely, that in the United States, Germany, and Japan. As a recent study documents, these regions require huge volumes of natural resources to support their industrial economies and the abundance of consumer goods and services they provide.[6] The U.S. economy, for example, annually consumes 20 metric tons (44,000 pounds) of natural resources—fuels, metals and minerals, food, and forest products—*per person*; Germany and Japan consume comparable amounts. With a few exceptions, virtually all of this material is eventually returned to the environment as waste or pollution. Nearly the only part of this process that intrudes on the broader public consciousness are the growing piles of municipal waste—the scrap heaps of throwaway products—and the shrinking space in which to put them. But that is a very small part of the total, less than three-quarters of a ton per person in the United States, and hardly the most serious in long-term environmental implications.

Additional environmental disturbance is required to extract resources and to build and maintain the roads and other infrastructure on which an industrial civilization depends. Americans, for example, annually move about 23 metric tons (50,710 pounds) of rock and dirt per person to get at the coal that provides much of the country's electricity. Also per person, agriculture and forestry erode 15 metric tons (33,070 pounds) of soil each year; mineral extraction shifts or processes 8 metric tons (17,640 pounds) of material; building of roads and maintenance of waterways moves another 14 metric tons (30,870 pounds). In all, the U.S. lifestyle requires the use of 80 metric tons of natural resources per person, every year; the total is the same in Germany and is about 44 metric tons (97,000) per person in Japan.[7] North America, large and well endowed, is nearly self-sufficient in raw materials; oil is the major exception. But for Europe and Japan, which import many of their natural resources, much of the environmental disturbance required to extract them occurs in other countries, often in developing regions.

The question is whether this is a workable model for a global industrial society, whether it can be scaled up without unacceptable environmental costs. A world of 9.5 billion people with, say, half of them consuming raw materials and generating pollution and waste at present U.S. levels would be a daunting prospect. But at present, that is the trajectory the world is on.

Consumption of energy and raw materials, now rising only slowly in industrial regions, is surging in developing regions. The unmet desire for energy, for a vast array of consumer goods, and for the basic infrastructure of an industrial civilization—roads, buildings, and utilities—is enormous. The number of cars, for example, is growing explosively: sales in China are expected to triple between 1990 and the year 2000.[8] Sales are also rising rapidly in Latin America. Indeed, by the year 2025, perhaps even sooner, the number of cars on the planet will double—reaching 1 billion—if current trends continue.[9]

In much of rapidly developing Asia, the demand for electric power was until recently increasing by 10 to 15 percent per year. China alone hopes to build about fifty new coal-fired power plants per year for the foreseeable future, in addition to the huge hydroelectric plant—the world's largest—planned for the Three Gorges Dam on the Yangtze River. The trend for energy use, in other words, is sharply upward. Moreover, consumers in developing regions have a long way to go before they catch up to industrial regions; the average person in China now uses only a tenth of the energy used by the average American; the average resident of India, only a thirtieth.

Energy production and use are major sources of air pollution. Other basic industrial activities, such as the production and use of chemicals and metals, are primary sources of more long-lived toxic pollutants. Just how much might these activities expand in the next half century? Studies of future energy demand, and detailed projections based on them suggest that energy use in developing countries may grow at least fivefold over the next fifty years, assuming that these countries' economies grow at only moderate rates. These projections also suggest that the production of chemicals and metals in developing countries could grow to roughly eight times present levels.[10] Projections for each developing region and the assumptions on which they are based are given in the appendix found at the end of this book.

What are the environmental implications of such a trajectory? If, for example, energy use in China expands by a factor of six over the next fifty years, so might emissions of pollutants, putting severe additional pressure on air quality. Similarly, if Chinese production of chemicals and metals expands tenfold, so might emissions of long-lived toxic pollutants. Whether air pollution and human exposure to toxic substances will actually increase as much as these projections suggest depends, of course, on what the Chinese authorities do—on how regulations are enforced, on how much is invested in clean technologies. That in turn may depend on whether China's government becomes more democratic and responsive to its citizens' concerns—in short, on what overall trajectory or scenario China follows into the future. But the projections nonetheless delineate how much greater the pollution pressures could become, if current practices continue, in each developing region.[11]

High levels of pollution can be costly, both economically and in human well-being. Factories and families alike must clean up dirty water before they can use it safely. In India, for example, many middle-class families boil tap water twice before using it to make tea; poorer families often can't afford to. Until recently, Mexico City's use of lead-ed gasoline and its notorious temperature inversions, which trap pollutants in the air for weeks, led to blood levels of lead in most children high enough to slow mental development; the U.S. Department of State advised diplomatic personnel against taking children to a post there. The World Bank estimates that more than a quarter of a million people die in China every year from pollution-related illnesses and that the country loses 3 to 8 percent of its annual economic output because of pollution and other forms of environmental degradation.[12] In Russia—perhaps the clearest example, along with eastern Europe, of the costs of unchecked industrial pollution—a survey shortly after the fall of the Soviet Union found nearly a third of Russian children afflicted with chronic asthma, thought to be caused by environmental contaminants. Health experts attribute Russia's plunging lifespans in part to widespread exposure to toxic materials.

Such a human toll may not directly constrain a country's destiny, but it is felt nonetheless. And since the burden of pollution usually falls most heavily on the poor, especially the urban poor, it becomes another barrier to economic and social development. One study found that inhabitants of urban slums in Latin America and Asia, unable to use polluted natural sources of water, paid 10 or even 100 times as much for water from a truck as did middle-class urbanites served by the municipal water service.[13] And pollution can become a potent political issue. Some experts believe that the 1986 nuclear reactor explosion at the Chernobyl Power Plant in what is now Ukraine, which riveted public attention on the Soviet Union and aroused widespread disgust with Soviet mismanagement of the disaster, was for many Russians, and especially for many Ukrainians the final straw, galvanizing support for political change and contributing to the collapse of the Soviet empire. Health experts worry that similar environmental health disasters may be brewing in many developing countries. Examples are the Brazilian gold-mining districts, where the mercury used to refine ore is usually dumped into rivers, the fish from which are the primary food supply for downstream communities, or in China, where the extraordinary levels of toxic chemicals pouring into some rivers may cause an epidemic of cancers in those who live along their banks.

Such hidden health and economic costs constitute one kind of indirect effect of industrial pollution. The potential for global climate change, driven by otherwise harmless emissions from the combustion of everyday fuels, is another. Here, projections of continued growth in worldwide energy use frame the central dilemma: energy use is essential to industrialization, but it is also altering the composition of Earth's atmosphere.

Several independent studies project that global energy use, driven by rapid expansion in developing regions, will expand by at least a factor of two and perhaps by a factor of three over the next half century.[14] Rising demand for energy is likely to be met by fossil fuels such as coal, oil, and natural gas—by far the cheapest and most convenient fuels for most uses—if current industrial patterns prevail. But burning of these fuels releases carbon dioxide, the most important of the greenhouse gases, which absorb infrared radiation and trap heat in Earth's atmosphere. Thus, if carbon dioxide emissions double or triple as they are projected to do if the current trends continue, we are likely to leave our descendants a significantly altered climate—one that could take several centuries or more to recover.[15]

There is now little doubt that the climate is beginning to change. A seminal international study involving hundreds of scientists recently concluded that "the balance of evidence suggests that human activities have already had a discernible influence on global climate."[16] Although uncertainties about the magnitude and timing of the change remain, models of Earth's atmospheric circulation and studies of the climatic record shed some light on the potential regional effects of climate change.[17] For example, sub-Saharan Africa and southern Europe are likely to experience more severe drought cycles, whereas middle-latitude regions such as the United States are likely to be subject to more flooding and heat waves. Overall, the climate is likely to be more variable, with more intense rainfall, longer periods of drought, and more intense heat waves. Although most industrial societies will probably adapt to climate change, the costs of doing so may be high—building seawalls or dikes in low-lying areas and coastal cities, abandoning farms and whole towns in floodplains, coping with an influx of tropical disease–carrying insects in the southern parts of the United States.[18] Developing regions, especially those still heavily dependent on subsistence agriculture, are likely to be more seriously affected. Populations in these regions may be forced to abandon low-lying areas and may face crop failures and possible famines from extended episodes of drought, with large populations vulnerable to more intense floods and storms. Moreover, warmer temperatures in tropical regions may lower crop yields by 10 to 20 percent. Island states such as the Maldives could lose large portions of their inhabitable land to rising sea, and a few could become entirely submerged.

Still more sweeping changes cannot be ruled out. If the currents in the Atlantic Ocean that bring warm surface waters northward and thus help keep the European climate moderate were disrupted by warmer temperatures and increased precipitation in high-latitude regions, as some scientists have speculated may happen, then Europe could be plunged into a mini–Ice Age. If the El Niño fluctuations that periodically bring torrential rains to the western coasts of the United States and South America and droughts to Australia and even South Africa, were to become much stronger—as evidence suggests has occurred in the past—many more countries would suffer. The rate at which temperatures may rise over the coming century is faster than that of any natural change documented in the past several hundred million years of Earth's geological record—a prospective change that the distinguished scientist Roger Revelle likened to a "planetary experiment" with human societies as the guinea pigs.

To stop the experiment and alter the energy trend that drives it is not impossible; nor would it necessarily mean halting the worldwide drive toward industrialization and urban lifestyles. But it would take radical changes in industrial patterns, requiring novel economic arrangements and new levels of political commitment—far beyond what a *Market World* scenario would entail.

Rural Impoverishment

Away from the burgeoning cities and rapidly growing industrial zones of developing countries, there is another world, one whose future also matters. More than half the world's population, 3 billion people, still live in rural areas—predominantly in Africa and Asia—in villages, isolated farms, and forest communities or as nomads following their livestock herds.

What most distinctly characterizes this portion of humanity is their near total dependence on the immediate environment for most of life's necessities, including water, food, cooking fuel, and forage for livestock.[19] A leading Indian ecologist, Madhav Gadgil, has used the term "ecosystem peoples" to describe them.[20] For these people, meeting basic needs is the primary activity, the central theme in life.

But their very dependence on the land has brought the world's ecosystem peoples to the brink of disaster. Poor farmers are forced to clear steep hillsides to farm them, knowing the land will erode in a year or two, because they have no other land and no other way to feed their families. Yet such acts only increase the difficulties faced by rural people, deepening their poverty.

Indeed, the three most basic natural resources—soil, forests, and water—on which ecosystem peoples depend are the ones at greatest risk of permanent degradation. Fertile soil is self-renewing when properly managed. But the world's stock of soils is being rapidly eroded, depleted of nutrients, and damaged by overgrazing and poor irrigation techniques. Over the past 50 years, 1.2 billion hectares, or nearly three billion acres of land—nearly 11 percent of Earth's soil resources, an area larger than India and China together—have been significantly degraded, lowering the soil's fertility.[21] In the eastern hills of Nepal, for example, fields abandoned because the topsoil washed away account for 38 percent of the land.[22] An estimated 6 million hectares (almost 15 million acres) of productive land becomes desert every year.[23] The extent of degradation varies by region but includes nearly a quarter of the land covered by vegetation in Europe, Central America, and Africa.

At the same time, the world has lost almost half its original forest cover, much of that within the past fifty years, and forests in developing countries are under enormous pressure. Deforestation—mostly clearing of forestland for agriculture—claimed more than 98 million hectares (240 million acres) between 1980 and 1990, an area nearly one and a half times the size of Texas.[24] Logging, often carelessly and destructively done, has decimated most of the remaining natural forests in Asia and many in Africa, and logging companies are now moving into South America's Amazon basin in large numbers.[25] Shifting cultivation—the ancient practice of cutting and burning a patch of forest, growing a few crops, and then moving on and letting the forest regrow—can no longer be continued. The presence of too many people in too little forest has shortened the fallow period so that forests fail to recover between cuttings; a recent international study found that shifting cultivation now destroys or seriously degrades 10 million hectares (25 million acres) of tropical forests every year.[26]

Water supplies are also threatened in many parts of the world. As forests are cleared in watersheds, stream runoff declines. Overpumping of underground aquifers is widespread in parts of China, India, Mexico, central Asia, North Africa, and the Middle East and in the grain belt of the United States. In Saudi Arabia, an extremely arid and water-scarce country, 90 percent of the country's water supply comes from fossil reservoirs created thousands of years ago; if current trends continue, the water will be gone within the first decade of the new century.[27] And even where water is abundant, pollution from untreated sewage or industrial wastes often renders it unusable.

Destruction of soils and forests, overhunting of game and overfishing in lakes and estuaries, and degradation of other critical renewable resources such as water are bad enough because they threaten to make these resources scarcer. But what makes the situation even more ominous are two other trends: continued population growth, which means that land and other natural resources must be stretched further, and economic competition between rural communities and cities for increasingly scarce resources, a competition the rural poor cannot win.

Research by economist Partha Dasgupta and other scholars has shown that for many poor families, the decision to have many children is usually a rational one.[28] Children (and sometimes cattle or other livestock) are among the few resources landless people control and are virtually the only resource poor families can increase without spending money. For poor families, children represent free labor, contributing to the family's livelihood by gathering wood, herding cattle, or, among the urban poor, working for wages in a store, a factory, or another family's home. For example, a recent study estimated that 50 percent of Indian children aged six to fourteen work.[29] Children, especially sons, often represent the only means of support for the parents in old age. And in many cultures, children are a source of status—proof of fertility for women and virility for men. Thus, large families are common among the poor, especially since parents know that some children are likely to die while they are still young.

And so population growth creates what has been called the downward spiral. As the population in a village grows, cropland, grazing land, and local fishing grounds all have to be shared with more people, leading to scarcity and often greater poverty. As the growing ranks of the poor struggle to survive, they are forced to farm more erodible land, cut down more forests, and crowd more livestock onto overburdened pastures, leading to degradation, still greater scarcity, and deeper poverty.[30] And poverty, as argued earlier, provides no incentive to have fewer children.

I have oversimplified this cycle of population growth, resource scarcity and degradation, impoverishment and more population growth, but I have done so to emphasize the interaction of demographic, economic, environmental, and sometimes social factors that is central to rural impoverishment. Among other implications, this interplay among factors means that there is little hope of conserving forests and other critical renewable resources in developing regions without also largely eliminating rural poverty—and that successful development in most poor regions cannot occur without halting rural impoverishment. Yet the trends are not promising. Consider how growing scarcity of these critical renewable resources might constrain development over the next half century.

One way to gauge potential scarcity is to assess a country's stock of fertile land per capita. When that stock falls below 0.07 hectare (about 0.17 acre) per person—the amount estimated to be needed to raise one person's food for a nonmeat diet, without modern methods of intensive cultivation, synthetic fertilizers, and pesticides—warning signals are indicated.[31] In 1990, not a single developing country fell below this threshold, although quite a few failed to feed their populations adequately. By the year 2025, however, 918 million to 3 billion people will be living in potentially land-scarce countries.[32] By 2050, 1.6 billion to 5.5 billion people—one-third to one-half of the projected world population—will be facing potential land scarcity, constraining development in India, China, and many countries in sub-Saharan Africa.

Similar estimates can be made for potential scarcity of water. Human use of freshwater has grown by a factor of four over the past fifty years, and a recent U.N. study projects that water use will increase by at least an additional 40 percent by the year 2025.[33] Supplies are already below the critical level—estimated to be about 1,000 cubic meters (264,200 gallons) per person per year to meet agricultural, industrial, and domestic needs—in about twenty countries with a population of 130 million people.[34] By the year 2025, 800 million to 1.1 billion people will be living in potentially water-scarce countries, and by 2050, the numbers will swell to 1 billion to 2.4 billion.[35] Virtually every country in North Africa and the Middle East except Turkey will face very severe water scarcity, and scarcity may also constrain development in parts of sub-Saharan Africa, central Asia, India, and China. Potential scarcity of land and water are described in more detail in the appendix found in the back of this volume.

Almost as essential as water and food for most poor rural people is cooking fuel—usually wood, sometimes crop wastes and cattle dung.[36] Fuelwood is already in scarce supply in many parts of Africa and Asia. There are "firewood deserts"—areas denuded of wood—stretching for miles around many African and Indian cities; in parts of West Africa, soybeans often are not planted as a subsistence crop, despite their high nutritional value and ease of cultivation, because they take longer to cook than yams or cassava and hence require more fuelwood.

Such constraints do not mean that developing countries will necessarily face actual shortages—land *can* be farmed more intensively, and water *can* be cleaned up and recycled, *if* these societies assemble the managerial and technical skills and take other steps necessary to modernize agriculture and water systems. But the constraints are real.

Adding to resource scarcities from rising population growth and from degradation of resources is a third trend: growing competition for such resources between rural and urban areas. Poor farmers usually lack the political influence to prevent diversion of water to urban areas, sale of logging rights to international companies, and invasion of traditional fishing grounds by the mechanized fleets that serve urban markets, even when the result may decimate resources vital to their livelihoods. Whatever the cause, scarcity of such renewable resources could undermine rural economies and further impoverish rural communities.

What are the prospects for rural populations, specifically the so-called ecosystem peoples? Some will succeed in creating a better life, especially in those countries that make a priority of rural development and agricultural modernization.[37] Others will find themselves in a downward spiral, increasingly unable to meet basic needs. Some may starve or endure chronic malnutrition. Many may choose to leave the land and seek their fortunes in already crowded urban areas. Some, increasingly pinched to obtain life's basic necessities while their better-off neighbors grow wealthy, may fight back. And it is just such conditions that raise the possibility of a *Fortress World*, of a future in which the urban rich get richer and the rural poor get poorer, of societies increasingly divided against themselves, of a future ripe for conflict and violence.

The next half century may see the emergence of a global industrial civilization, one that could provide increased opportunities and more comfortable urban lifestyles for billions of people. But if current patterns of industrial activity persist, the result could also be a far more polluted planet, possibly even an irrevocably altered Earth. Might that constitute a kind of false prosperity in which our descendants are wealthier in economic terms but poorer in their environment and their quality of life? Moreover, industrialization is only one of the threats to the planet's ecological balance. Already, human activities have altered more than one-third of Earth's surface, consume more than 40 percent of its biological production, and threaten significant proportions of its animal and plant species with extinction.[38] If current trends are not changed, rural impoverishment and rampant demand to support urban lifestyles could so degrade soils, forests, fisheries, and other productive ecosystems as to decimate a large part of the heritage built up by eons of evolution—leaving poor countries bereft of the very resources they need for development and leaving to all future generations a biologically impoverished planet.

Chapter 8

Critical Security Trends

WILL THE WORLD OF THE TWENTY-FIRST CENTURY be safer or more threatening? The answer matters not only because personal safety weighs heavily on all of us but also because progress—economic, social, and environmental—requires stable societies. And as the world becomes ever more interconnected, instability in one place can have devastating effects elsewhere.

Frightening images come to mind. If rising regional tensions within China were to erupt into civil war, military officials in the Pentagon would stay on alert and the stocks of Western corporations with major investments in China would plummet. Conflict in the Middle East—whether spreading Islamic revolution or wars over water—would again threaten world oil supplies. Nuclear weapons or the materials to make them seized by Russian criminal organizations and sold to rogue states or terrorist groups would be security experts' nightmare come true.

Equally frightening is the prospect of increased threats to the personal security of individuals. Ethnic and religious conflicts appear to be on the rise and often erupt into spontaneous violence, killing large numbers of people and sometimes, as in Bosnia, destroying a country's economy. Government collapse leading to a state of lawlessness, banditry, and warlords is becoming increasingly common in parts of Africa. Smaller-scale social and criminal violence—from carjackings to church burnings to drug-related killings—touches individuals and communities in virtually every country. Terrorism plagues many regions of the world and reaches even the United States, Japan, and other industrial countries, threatening airplane passengers, subway riders, and office workers.

Such violence may not seriously threaten the state, but it does affect the lives, peace of mind, and behavior of individual citizens. For example, many wealthy families in Latin America and, increasingly, business executives from Moscow to Mexico City already hire bodyguards to protect themselves and their families from kidnapping; in Colombia, even middle-class families are becoming targets.[1] If severe enough, such threats can disrupt whole societies and undercut national well-being: they have the potential to shift public attitudes and shape new national priorities, diverting money from social services to fight crime, for example, or leading to repressive laws that subvert democracy.

People worry about more than violence, however; they also worry about losing their jobs and about new threats to their health. Immigration on a scale large enough to threaten jobs or the social balance, for example, could become politically and socially destabilizing. Increasing tourism and business travel puts society at risk of emerging diseases. For example, medical authorities point out that repeated outbreaks of the deadly Ebola virus, presently confined to Africa, are only about twenty-four hours and a plane ride away from virtually any major city on Earth. Public health officials fear that a warming climate could unleash a plethora of tropical diseases—such as mosquito-borne dengue fever, now spreading north from Central America—on the United States.

Such threats to economic and environmental security can have profound effects at the national level. For example, scarcity of land and growing economic desperation among peasant families in the Chiapas region of Mexico contributed to the 1994 Zapatista uprising and, possibly, to the subsequent international loss of confidence in the peso and

Mexico's resulting economic crisis.[2] For tourism-dependent Caribbean countries, dengue fever could become economically devastating: a high chance of catching a nasty disease would seriously undermine the appeal of sunny beaches.[3]

As the specter of nuclear war recedes and the chance of even conventional war declines—as now seems likely—these new security concerns are growing in importance, to the point at which they may increasingly threaten human progress in the coming century. Indeed, many analysts now argue that a profound change is occurring in the nature of the threats society faces.[4] If the twentieth century was dominated by war and the threat of external aggression, the twenty-first century may be defined by terrorism, crime, large-scale population movements, economic and environmental threats, and social instability. These emerging threats thus constitute a critical trend.

Could such stresses overwhelm societies and their governments; erode civilization's social, legal, and moral underpinnings; place ethnic and religious groups or rich and poor in direct conflict; even lead to anarchy? The future foreshadowed in *Fortress World* indeed raises the possibility that the coming century will be a far grimmer place.

Challenges to Stability

The emerging security threats will challenge the capacity of states to govern, of nature to provide, and of societies to cope more than they will test the mettle of armed forces. Against these kinds of threats—most of them internal—"smart" weapons and stealth bombers are irrelevant and military forces often ineffective. What external threats remain are mostly not of a conventional military nature. Thus, the security problems of the coming century will increasingly involve threats to personal security and to the stability of society, in rich and poor nations alike.

The wealthy industrial nations of the West may be spared the direct effects of some, but not all, of these novel security threats. In an increasingly integrated world, misery is often easily exported, and the economic and social costs of conflict and instability are likely to be widely shared.

What are the underlying causes of these new threats to stability and security? There is no single, simple answer because the list of threats is diverse and their causes range from the growing worldwide availability of dangerous technologies to the demographic and rural impoverishment trends described earlier. And although no single threat may appear overwhelming, several together may prove just that, especially in developing societies with fragile institutions. The following sections provide a quick overview of a number of these emerging security concerns to illustrate why stable societies cannot be taken for granted.

Globalization of Crime

Global businesses thrive on increasingly open borders, cheap global communications, and electronic financial transactions. But just as legitimate global corporations can site facilities and organize their financial matters to take advantage of national differences in tax laws and regulatory practices, so, too, can criminal organizations.

Major drug traffickers, for example, can grow their product in one country, refine it in another, distribute it through complex and rapidly changing global routes, and recycle their money through circuitous financial channels. National law enforcement efforts, especially in developing countries, are simply no match for the sophisticated drug cartels. Where law enforcement cannot be evaded, it can often be bought: the international drug business alone generates $500 billion per year, more than the entire national income of all but a handful of countries.

Competitive pressures simply make matters worse. Governments everywhere, responding to the needs of legitimate businesses, are streamlining inspections and border controls. U.S. customs officials at busy Mexican border crossings, for example, are told to spend no more than one minute examining each northbound truck, to avoid huge traffic jams.[5] Shipping of contraband—whether drugs, stolen antiquities, or weapons—is becoming ever easier.

At the same time, money laundering is rising. Large amounts of money move in and out of offshore banks, largely beyond the control of governments, allowing some businesses and wealthy individuals to evade taxes or hide corporate ownerships. These offshore institutions make it easy to recycle illegal profits or to pay the money that lubricates bribery and corruption. With more than a trillion dollars every day moving via computer through international cyberspace, "following the money" in illegal transactions is becoming increasingly hard.

The drug trade is a direct threat to citizens and legitimate business-es, but it can also undermine governments, distort judicial systems, and create a climate of corruption violence, threatening national destinies. President Ernesto Zedillo of Mexico has described narcotics trafficking as that country's greatest national security concern.[6] Moreover, crime both feeds on and contributes to instability. Drug traffickers are report-edly moving into Bosnia, exploiting the unsettled conditions there to transship drugs to the rest of Europe; Russia's inability to control extor-tion and other forms of organized crime discourages foreign investment in that country, thus delaying its economic recovery.

Proliferation of Dangerous Technologies

Imagine the shock and horror if a terrorist group were to set off a nuclear explosion in a U.S. city. Far-fetched? Yet the Central Intelli-gence Agency says that nuclear materials and the technology to build bombs with them are "more accessible than at any other time in histo-ry," and Harvard University professor Graham Allison, an expert on nuclear proliferation, is reported as saying that nuclear terrorism in the United States is a matter of not "if" but "when."[7]

Information about dangerous technologies is widely available. Two decades ago, a determined individual could assemble several file cabi-nets full of documents that were publicly available in the United States and that described in great detail every aspect of making fairly simple atomic bombs.[8] Information on guns, military weapons, and explosives is more widely available today—in magazines catering to hunters and military buffs and in catalogues and brochures from the gun industry, as well as in more specialized or purposeful sources. Detailed instruc-tions for making fertilizer and fuel oil bombs like that used in the Okla-homa City terrorist attack circulate as pamphlets and newsletters among U.S. right-wing militia groups, over electronic bulletin boards, and in Internet conferences.

Beyond such specific information, the underlying scientific knowledge and the core components and tools for many potentially dangerous technologies are widely available. The basic electronics required to make timing, ignition, and guidance devices exist in every industrial country; the open scientific literature contains all the knowledge needed to synthesize dangerous poisons from commercially available chemicals; the molecular biology skills needed to develop germ warfare agents are increasingly widely available. As the Japanese Aum sect responsible for the Tokyo subway attacks demonstrated, even determined private groups can assemble the materials and the skills required to make chemical or biological weapons of mass destruction.[9]

In addition, the risk that nuclear weapons might be detonated, either to gain advantage in a regional conflict or as a terrorist act, appears to be rising. Some 250,000 kilograms (551,000 pounds) of plutonium reside in military stockpiles—a troubling legacy of the cold war. With knowledge of bomb making widespread, the safeguarding of such nuclear materials is vital but difficult—especially in today's Russia—given that only a few kilograms (an amount about the size of a grapefruit) is enough to make a small atomic bomb and given the high price plutonium could command on the black market. With controls over Russian stockpiles dependent on the stability and effectiveness of that country's downsizing military establishment and with economic hardship widespread there, the potential for diversion—and for nuclear terrorism—is high.[10]

For the United States and other industrial nations, the most worrisome nuclear risk may be that a single bomb or the material to make one would fall into the hands of a terrorist group, foreign or domestic, determined to use it. Small enough to carry in a suitcase, the plutonium for a terrorist weapon would not be difficult to smuggle into the country. And, as terrible as the Oklahoma City bombing was, the destruction from a small nuclear bomb delivered in the same way would be thousands of times worse. Huge investments in national defense are not much help against such nonmilitary threats.

Conventional weapons are also proliferating. Military contractors, faced with downsizing defense budgets, are actively seeking markets for their tanks, fighter planes, rocket launchers, and other weapons, often with the support of their national governments. The United States is in fact the world's largest arms merchant, but arms makers in Europe, Russia, China, and other countries are also eager to sell. Quite apart from legitimate, nationally sanctioned arms sales, there is an active illegal international trade in weapons of all kinds. For a price, would-be guerrilla leaders and terrorist groups can obtain a wide variety of modern weapons, including, as past experience shows, rockets capable of bringing down an airliner or destroying a building.

As a result, today even a small nation or a dedicated private group can cause horrible destruction. High levels of spontaneous violence, hit-and-run urban terrorist attacks, even dispersed civil war can have a profound effect on the quality of life, on the climate for economic progress, and—in fragile nations—on national survival. The technological means for such violence already exist; what about the underlying motives for violence and civil disorder?

Unemployment and Migration

In 1996, unemployment in the European Union countries was at 11 percent, a worrisome level, with more than 20 million people out of work in western Europe as a whole.[11] Among young people, unemployment levels were far higher—nearly a quarter of French men and women under the age of twenty-five were without work; many of them had never had a job. Generous unemployment benefits keep people from hardship, but not from the social and psychological effects of chronic unemployment.[12] Most of the chronically unemployed are men, especially those lacking advanced skills. Is such a large pool of idle men a threat to the stability of European societies? At the very least, this cohort is a fertile recruiting ground for gangs, extremist political groups, and criminal activity. Certainly, chronic unemployment fuels crime in the U.S. inner cities and can cause instability, as past riots attest. Welfare reforms may initially add to such problems.

Demographic trends seem likely to make things worse. The aging of European societies is undermining the social contract that provides generous old age pensions and unlimited unemployment benefits because there will not be enough workers to support those receiving such benefits. In effect, European governments (and to a lesser extent, those of the United States and Japan as well) have made promises they cannot keep: benefits are likely to decline.[13] But decreasing or denying benefits to large numbers of unemployed people could cause a social and political explosion, especially in Europe. Even extremely modest attempts to tighten benefits in France brought a nationwide truckers' strike and rapid government capitulation.

Illegal immigration, fueled by rising gaps in income between rich societies and poor ones, may well add to such economic and social pressures. Indeed, immigration could surge if conditions in developing countries deteriorate suddenly—a radical Islamic takeover in Algeria, for example, could send many refugees to Europe; economic collapse in a Central American country would undoubtedly intensify the flow of people into the United States. In an apocalyptic novel, *The Camp of the Saints,* French author Jean Raspail evokes just such a future in which, gripped by poverty and hunger, more than a million people set out in an armada of small boats to find a better life, eventually coming ashore on the Mediterranean coast of France.[14]

But even without such dramatic developments, the inexorable pull of what seems to many poor people to be lands of incomparable wealth is real enough. Ask nearly any of the young men and women in the urban centers of the developing world whether they would like to go to the United States or to Europe, and the likely answer is yes. In societies in which the average wage is still only a few dollars per day, the images of U.S., European, and Japanese lifestyles that fill the media are powerful indeed. Individuals who think they can increase their earning power by, say, tenfold may well decide that the hazards of illegal migration are worth it. The flood of Mexicans and others moving illegally into the United States over the past decade, despite a more vigilant border patrol, suggests what the future may bring if income gaps continue to rise. Already, the smuggling of Chinese into the United States and of Africans and Asians into Europe is a lucrative business—more lucrative, by some accounts, than the smuggling of weapons or drugs.[15] Even if countries gain control over their borders, there is another way in. Anyone who can get a tourist or business visa can simply fly in, walk off the plane, and disappear—and despite screening of those who seek tourist visas from poor countries, control of this type of illegal entry is almost certainly a losing battle. Countries cannot stop illegal immigration without shutting off the tourist trade and closing their borders to commerce, which are not realistic options.

Around the world, resentment of immigrants, both legal and illegal, is rising, especially where their concentrations are high, as with Mexican immigrants in southern California, and where cultural differences and competition for jobs are major issues, as with North African immigrants in France. In recent decades, an estimated 12 to 17 million people from Bangladesh have poured into the Indian states of Assam, Tripura, and West Bengal, changing landownership patterns and the balance of political power and setting off several instances of ethnic violence.[16] And people from across the African continent are moving to South Africa, adding to that country's social problems.

Pressures from unemployment and migration could turn industrial societies inward, bring to power reactionary governments, and support the adoption of repressive laws and protectionist economic measures—in short, drastically change the nature of Western societies. That may not happen, but concerns about economic and social insecurity—and social instability—are not likely to subside.

Rural Impoverishment and Instability

The rural poor constitute one-third to one-half of humanity. Most of them are concentrated in Africa and South Asia, but they are also spread throughout Latin America, Southeast Asia, and China. Nearly all of them depend on natural resources such as soils, water, forests, and grasslands for their livelihoods. Yet not only are these resources becoming rapidly degraded, they are also increasingly being diverted to higher-paying and politically more powerful users, leaving rural people to make do without. Rising populations are adding to these pressures. Together, these trends threaten to create a resource crisis for rural people, impoverishing them further.

When people are squeezed too far, when they can no longer sustain themselves, they usually go somewhere else—to frontier areas where there are unexploited resources (if such places exist) or, increasingly, to cities. But sometimes people resist being squeezed, or they attempt to fight back at those they perceive as responsible. In the Chiapas region of Mexico, for example, the population doubled between 1970 and 1990, and that of the poorest social group, the native peasants known as *indigenas,* tripled. Land suitable for farming became increasingly scarce; cultivated land per capita has been declining sharply since 1980. *Indigenas* seeking to create new land by clearing forests caused new problems: shortages of firewood, water holes that dried up, and soil erosion. Fierce competition for land among commercial farmers, cattle ranchers, loggers, and *indigena* communities often led to conflict, with the *indigenas* usually being evicted. What proved the last straw for increasingly radical peasant groups was a 1992 constitutional amendment that removed protections for community-owned land.[17] The Zapatista National Liberation Army was formed, arms and supplies acquired, and a wide group of peasants mobilized in support. In early 1994, the Zapatistas seized and briefly held a number of cities in the region, dramatically bringing the plight of Chiapas's peasants to the world's attention.

The uprising, as a recent analysis suggests, illustrates the interaction of resource scarcity, rural impoverishment, and conflict.[18] Nor is Chiapas the only such example. Guerrilla attacks in the Philippines, the *Sendero Luminoso* (Shining Path) rebellion in Peru, and the murderous ethnic conflict in Rwanda—which, at less than 0.33 hectares (1 acre) of land per person, has the highest population density in continental Africa—in all these examples, resource scarcity and rural impoverishment have played a role.[19] And in coming decades, such pressures will confront dozens of countries, including many in Africa and parts of both India and China. Will their governments be up to the challenge, or will rural instability and related social conflicts become increasingly common?

Resource scarcity can also become a direct cause of conflict. The clearest examples come from water shortages, which are critical in much of the Middle East. Some analysts believe that Israel's expropriation of underground water supplies from the occupied West Bank of the Jordan River, whereby it denied or restricted the use of these aquifers to Arab farmers in the region, contributed to the grievances that led to the Intifada uprising against Israel in 1987–1993.[20] Nearly 40 percent of the groundwater Israel now uses comes from occupied territory. Tensions over water are likely to get worse because Israel's water tables are dropping due to overuse even as water needs in Israel, the West Bank, and Jordan are growing. Israel's demand alone may outstrip supply by 40 percent within three decades.[21] Similar potential conflicts await Egypt, which already uses all the available water from the Nile River, and upstream countries such as Ethiopia and Sudan. Turkey controls the source of the Euphrates River and is building an extensive series of dams, yet the river is also critical to both Syria and Iraq. Wars over water cannot be ruled out.

Urban Unrest

The world's urban population is growing by more than 1 million every week, mostly from massive rural-to-urban migration. Urbanization may be an inevitable consequence of the modernization of agriculture, but today's migrations are exacerbated by growing rural hardships and the prospects of better jobs and higher wages. In China, for example, urban incomes are three times rural ones and are rising twice as fast. More than anything, however, it is the unprecedented scale of expected urban growth that staggers the imagination. China's urban population, for example, is expected to swell by more than 300 million over the next fifteen years.[22]

Worldwide, urban populations are expected to rise to 2.5 billion by the year 2025, nearly all in developing countries.[23] The number of cities with populations greater than 1 million is expected to swell from 290 in 1990 to 550 or more by 2015. Some of these cities will reach sizes beyond all human experience, dwarfing even today's megacities, placing enormous strains on surrounding natural resources, and challenging human capacity to manage them.

The informal character of urban shantytowns creates additional problems. Without legal title to the land, residents hesitate to build permanent houses or to invest in such community services as garbage collection and sewage systems. As a result, most shantytowns are badly polluted and unhealthy places to live.[24] Many also harbor a host of social problems, including frequent violence and crime. Some of the well-established shantytowns around Rio de Janeiro are now ruled by drug lords. In China, urban immigrants—even those with work permits—have no legal right to settle permanently in urban areas and consequently are not entitled to health services, transportation services, or emergency food supplies; yet nearly 40 percent of the population of Tianjin, a major industrial city, is composed of such illegal immigrants.

What happens if, say, the Chinese economy should hit a major downturn and cities refuse emergency food supplies to millions of suddenly unemployed workers? Might these people stage protest marches demanding food and other rights or turn to looting and riots? At the very least, such huge and marginal urban populations are a potential source of instability in China, in Africa, and in India—as they already are in Latin America.

Inequity and Instability

Certainly, urban areas exhibit the extremes of economic disparity, since they bring together both the wealthiest parts of a society and, increasingly, some of the poorest. Could such disparities, if they become too extreme, undermine the stability of a society?

Latin America, which has the most extreme economic disparities of any region, is in many ways a laboratory for such concerns. Some observers argue that the failure of the region's governments to alleviate severe poverty, despite improving economic conditions, threatens to undermine democratic reforms.[25] The recent U.S. ambassador to Mexico, James R. Jones, compared the situation in Mexico to that during the Great Depression in the United States, estimating that about half of the Mexican population is not yet benefiting from economic reform. "Opportunities have got to become real for these people," he said. "Otherwise [they] are going to lose faith in free markets and democratic freedoms. I think we're living on limited time."[26]

Poverty and inequity by themselves may not directly elicit conflict. But in a grossly inequitable society in which the poor lack much hope for betterment, volatile conditions can arise and isolated events can trigger violence. Examples include urban riots in the United States sparked by events such as the slaying of Martin Luther King Jr. or the beating by police of Rodney King. Likewise, smoldering grievances, when combined with perceived loss of government legitimacy, can be transformed into active opposition and violence, as in the anti-apartheid movement in South Africa. In short, widening economic disparities within societies heighten the potential for instability and violence.

———

The impact of emerging security problems could be immense. In direct security terms, widespread social collapse—ten Bosnias or Rwandas or Haitis at once—could rapidly become unmanageable, producing huge movements of desperate populations. In human terms, it would be hard to ignore tens of millions of deaths from slaughter, starvation, or disease among refugees—and yet human casualties on a scale to dwarf the Holocaust are entirely possible in the coming century. Misery and tragedy on such an unprecedented scale would conflict with any notion that humankind had at last achieved a mature, global civilization.

Such dismal futures, consistent with a *Fortress World,* need not happen. But to avoid them means addressing the basic causes of instability everywhere—accepting the obligation to be our neighbors' keeper because, ultimately, our neighbors' problems will become ours. This obligation is usually posed on purely moral grounds; here, however, I also argue that pragmatic self-interest in our personal security and in the stability of our societies leads to the same conclusion.

Chapter 9

Critical Social and Political Trends

WHEN SOCIAL HISTORIANS many generations hence chronicle the twentieth century, they may single out the rise in status of women as perhaps its most important feature. Nearly everywhere, women are better educated than they were even a few decades ago, with more access to medical care and to family planning services and more control over their fertility. The trend is still far from universal—the lot of many poor women has not improved and progress has been very limited in some regions—but nonetheless, there has been progress. Women are also playing a more active economic role, working increasingly outside the home in factories, offices, and retail stores and in the informal economies of many developing countries. And, in a few countries, women are emerging as a real political force.

Although these changes are occurring slowly, they are inexorably altering the status of half the human race—certainly a critical trend and a hopeful one. Indeed, most of the social and political trends I mention in this chapter are hopeful, unlike many of the other critical trends examined so far. Can these hopeful trends counteract or reverse pessimistic demographic, environmental, and security trends? Not by themselves. But as we will see, they may represent opportunities for human societies to control their own destinies—to create a *Transformed World* and a more optimistic future.

The Rise of Women

The rise in the status of women is a worldwide phenomenon, a trend that has gone furthest in Scandinavia and the United States. Consider that before 1917, women in the United States had no right to vote and virtually no economic power independent of their husbands. Yet in the most recent U.S. election, women not only voted in larger numbers than men but also, in many cases, voted differently from their husbands, providing the decisive margin in reelecting President Bill Clinton. And in recent decades, businesses owned and managed by women have grown in number so rapidly as to now constitute nearly a third of all U.S. businesses.

If these trends continue and spread globally, they are likely to have a profound effect on the next century. Women may alter the political agenda, giving higher priority to social issues, or may be less willing for their nation to engage in war. Their economic power may shift consumer preferences. And their increased participation in the workforce will accelerate the spread of prosperity—indeed, may be crucial to national economic success.[1]

Demographically, women's educational level plays an important role in family size. Data from dozens of developing countries show a strong decline in the number of births per woman as female literacy rises.[2] Moreover, women in such countries who find jobs outside the home and so have income independent of their husbands tend to spend most of their money on the health and education of their families, whereas men tend to spend most of their income on themselves.[3] So the rise in the status of women will play an important role in slowing population growth, in eliminating poverty (since most of the poor are women), and in improving social development.

One indication of how rapidly women in developing regions are gaining control over their own fertility is the spread of family planning practices. In the 1960s, less than 10 percent of married couples in developing countries used contraceptives; in the 1990s, more than 50 percent do, a virtual revolution.[4]

But even though the trend is clear, the overall transformation of women's status is far from complete. The macho culture in Latin American societies, the traditional dominance of males in most African and Asian societies, and the restrictive sanctions on women's activities outside the home in many Islamic cultures all create barriers to the rise of women. Of the world's nearly 1 billion illiterate adults, two-thirds are women; 70 percent of the 130 million children of primary school age who are not enrolled in school are girls.[5] Many women cannot control their destinies; some live in virtual slavery. In developing countries, a third of all women are battered by their husbands, according to one report, and an estimated 100 million girls, mostly in Africa, have suffered genital mutilation.[6]

How rapidly the status of women improves may be critical for the fortunes of some of the poorest developing countries. Still, economic necessity is a powerful incentive: even a conservative Islamic country such as Saudi Arabia, faced with either letting women work or importing more foreign workers (who are feared as a source of potential instability), has chosen to increase women's economic opportunities. In an increasingly global economy, countries that do not make use of the abilities of half of their adult population will fall behind.

Human and Social Development

The status of women is one critical element of human societies, but what about broader trends in social development? Here, too, the trends are strongly positive. A series of reports by a U.N. agency show that developing countries have made huge strides in human development over the past thirty years—progress that took a century in the industrial countries.[7] For example, enrollments in primary and secondary schools has more than doubled, infant mortality rates have fallen by half, and people in developing countries now live seventeen years longer on average than they did three decades ago.

But these overall trends disguise the fact that progress from region to region remains very uneven. Sub-Saharan Africa, for example, lags behind most other regions in education, in life spans, and in access to health care; the region has only one doctor for every 18,000 people. Moreover, social and economic development do not always march in lockstep: the wealthy Arab states lag behind in female literacy, for example, with nearly 60 percent of women in those countries still illiterate despite generally high incomes. In South Asia, even more people—more than 60 percent of the population—are deprived of basic social services than are poor in the sense of low incomes.[8]

One way to visualize the future is to ask how long it would take each developing region to reach human development levels comparable to those in the industrial countries today if recent trends continue. If China continues along its current path, for example, its population will reach modern levels of social development within just twenty-five years, according to U.N. projections. India will take more than a century to reach the same level; much of Africa will take more than two centuries.[9] If the pace of social development does not accelerate, other critical trends such as rising populations, growing resource scarcity and rural impoverishment, and gathering forces of anarchy may overwhelm human development.

Improvements in basic social services such as education and health care are not the only concern: the quality of a nation's social fabric is important, too, and there are signs that it is fraying worldwide, even where social development is well advanced. The basic unit of society, the family, is showing signs of stress, and even what constitutes a family unit is changing. In the United States, for example, nearly a quarter of all households with children are headed by a single parent, representing a sharp increase in this phenomenon since the early 1970s.[10] In Europe and in a number of developing countries, the numbers are lower but are rising rapidly, fueled by increasing numbers of divorces and births to unwed mothers.[11]

Divorce rates have doubled since 1970 in many industrial countries. In many countries, close to a quarter of all marriages end in divorce; in countries ranging from the United States to the Dominican Republic to Ghana to Indonesia, divorce affects nearly one marriage in two. At the same time, childbirth outside marriage accounts for nearly 20 percent of births in Europe as a whole and nearly 40 percent in France and the United States. Although unwed mothers are less common in Islamic countries, they are becoming more common in other developing regions, especially Africa. In Botswana, for example, 40 percent of unmarried women have a child by the age of twenty.

The consequences for children of having a single parent can be severe. In many developing countries, such children are at far greater risk of malnutrition and may never attend school regularly because their mothers lack the cash to pay school fees. Even in industrial countries, single-parent families are far more likely to be poor, and the children are at far greater risk of never completing school, of becoming involved in crime, and of facing poor economic prospects as adults.

In parts of Africa and increasingly in parts of Asia and Latin America as well, AIDS adds to the stresses on families. Most of those who die from the disease are in their prime childbearing years; as a result, the number of children orphaned by AIDS is rising very rapidly. Zimbabwe, in southern Africa, counted some 65,000 AIDS orphans in 1993 and expects more than 600,000 by the year 2000.[12] Although orphaned children in much of Africa are traditionally absorbed into extended family networks, these risk being overwhelmed. More and more, elderly Africans—nearly always grandmothers—are finding themselves without the support of their children in old age and facing the additional burden of attempting to care for their grandchildren. In Uganda, so many children have lost both parents and often other relatives as well that relief workers use the term "child-headed households" to describe the growing and tragic phenomenon of children trying to cope entirely on their own. Lack of adequate social services—and in many communities, a stigma attached to the children of those with AIDS— means that many orphaned children face a difficult future.

Cultures at Risk

To some degree, every developing culture is threatened by exposure to the industrial world. Global communications media, rising tourism, and the growing presence of advertising by multinational corporations ("Drink Coke!" "Smoke Marlboro!") guarantees that images of wealth and the consumer products of modern industrial countries are inescapable. Especially the youth in many cultures, anthropological studies show, are susceptible to these images: they not only desire sunglasses, jeans, Walkmans, and other symbols of Western lifestyles but also, all too often, come to despise or feel ashamed of their traditional culture.[13]

What the youth in developing regions aspire to is important, in part because there are so many of them. The influential baby boom generation in the West was just a single generation, but it remains an ongoing phenomenon in most developing countries, where a large proportion of the population is under the age of twenty. There will be more than 2 billion teenagers in the world at the turn of the century, mostly in developing countries, and many of them seem to share not only a fascination for jeans and rock music but also a restlessness for change and other adolescent characteristics that make them a volatile social force—a phenomenon futurist Peter Schwartz has dubbed the "global teenager."[14] These young people will be the primary drivers of social and cultural change in their countries.

But the cultural assault goes beyond consumer products. Learning English is increasingly necessary for aspiring scientists, software developers, retail clerks in tourist shops, and international business executives, among others. The spread of English is being enhanced by the Internet and other rapidly growing electronic communications media and by the U.S. dominance in movies and software. Within a decade, according to one study, more people will speak English as a second language than as a mother tongue.[15] Some see the growing dominance of English as a cultural threat to other languages and ways of life. Indeed, the president of France, Jacques Chirac, has called the spread of English "a major risk for humanity."[16]

France is hardly alone in its concerns. Leaders in some Asian countries and a number of Islamic countries have described Western lifestyles as decadent and Western ideas as the seeds of moral decline—in effect, as threats to their cultures.[17]

Clearly, as the world becomes a smaller place, cultures are placed under growing stress. Many societies may find it impossible to continue traditional practices; modernization, in one form or another, may be inevitable. Although the loss of traditional cultures represents a blow to human diversity, some changes—such as the roles they assign or permit to women—may be for the better, an essential step toward stable populations and away from poverty.

The Changing Nature of Governance

Worldwide, dissatisfaction with governments is running high and debates about their role and effectiveness are widespread. Certainly, governments no longer have a monopoly on power—other powerful institutions are emerging, and the structure of authority in many societies is changing. Governments, in short, are less able to dominate events, yet they face more complex problems.

Consider a few of the challenges to national governments posed by a globalizing world. On any given day, nearly $1.25 trillion flows through global capital markets, an amount larger than the *annual* economic output of all but a few countries. As a result, no country's economic policies are entirely sovereign anymore but are influenced by a market that can force changes in interest rates and alter the value of currencies—sometimes drastically, as Mexico and, more recently, many southeast Asian countries have found, to their woe. In addition, the economic power of global corporations increasingly rivals that of the governments of small and even medium-sized countries and so can affect a country's economic fortunes. Corporate decisions can create or destroy jobs, affect the exploitation of natural resources, and shape the destinies of communities. Furthermore, global communications systems, from CNN to telephones to the Internet, mean that even authoritarian governments can no longer control the information available to their citizens, although many still try. "Increasingly, resources and threats that matter, including money, information, pollution and popular culture, circulate and shape lives and economies with little regard for political boundaries" is how Jessica T. Mathews, president of the Carnegie Endowment for International Peace, sums up the loss of government power and autonomy. "Even the most powerful states," she argues, "find the marketplace and international public opinion compelling them more often to follow a particular course."[18]

Even as governments' control wanes, however, a new and improbable entrant in the sweepstakes of power is gaining influence. Its components are a wildly diverse stew of nongovernmental organizations and entities that take many forms, from citizen's and community groups to religious organizations to activist groups espousing various environmental and social goals. Their strength lies partly in their intensity of purpose, fueled by volunteers and donors who support their cause; partly in their credibility as not-for-profit or grassroots organizations, serving a larger social goal; and partly in their sheer numbers. Mathews estimates that there are millions of such citizen's groups, ranging from "the tiniest village association to influential but modestly funded international groups like Amnesty International to larger global activist organizations like Greenpeace International and giant service providers like CARE, which has an annual budget of nearly $400 million."[19] Working alone or, more frequently, in alliances or informal coalitions, such citizen's groups have become increasingly able to focus public opinion, shape political agendas, and even help frame international laws, such as the 1997 treaty banning land mines. In many developing countries and in the urban slums and troubled neighborhoods of industrial countries, such groups provide emergency relief, operate soup kitchens and homeless shelters, and provide an enormous range of social and development services, from running drug rehabilitation programs to teaching better farming techniques to helping communities form weaving cooperatives. In effect, citizen's groups—sometimes called "civil society"—are increasingly filling voids created by governments' inaction or incapacity, finding novel and often inexpensive ways to meet neglected social needs, and giving a voice to overlooked causes and powerless segments of society. The rise of civil society—like that of global markets and global media—is altering the process of governance and so may be a hopeful lever for change.

But there is a still more fundamental trend in the process of governance, and that is the spread of democracy around the globe and the democratizing of government processes, from school board hearings to urban planning processes to U.N. meetings. Nearly all of Latin America and a growing portion of Africa and Asia now have democratic governments, even though some are extremely fragile and a number are democratic only in outward form; China has introduced democracy at the village level, selecting local officials through free elections in nearly a million communities. In most industrial countries and in some developing regions, governments are encouraging increased citizen participation in administrative hearings and community planning processes, not just elections. In the United States, many county government meetings and judicial proceedings are televised to provide wider access, sometimes, as in the O. J. Simpson trial, commanding a national audience. U.N. negotiating sessions—until recently for governments only— now include representatives of citizen's groups, giving these interests a direct role in international governance.

The result of growing democratization, in the long run, will be more responsive governments. Although there are instances of ineffective democratic governments and competent authoritarian ones—as demonstrated by India's poor track record in managing development over the past fifty years and the success with which China's government has guided that country's rapid economic and social rise—a far larger number of authoritarian governments have squandered their countries' resources, mismanaged their economies, and sometimes stolen their wealth, from the Philippines under Ferdinand Marcos to Nigeria's current military government. If free elections and open government processes are no guarantee of good government, they appear nonetheless to be the best available strategy.

Moreover, economic necessity suggests that even successful authoritarian governments such as China's may eventually be forced to share their power. Modern market-based economies rely on widespread initiative, independence of judgment and action, and, in a global economy, the ability to interact with a wide variety of foreign individuals and organizations—all of which are severely hampered under an authoritarian government that does not trust its citizens and tries to limit their freedom. Without democratic governance, countries may lose some of their most talented people to opportunities elsewhere—as exemplified in the extreme by the exodus of doctors, teachers, and businessmen when the Taliban Islamic Movement took over Kabul, the capital of

Afghanistan, and began to impose its rigid, fundamentalist version of Islam; such countries are likely to lag behind more open societies in development.

In addition, democracies tend to tolerate the activities of citizen's groups, whereas authoritarian regimes tend to forbid or restrict them. Such organizations are noticeably rare in regions such as China and North Africa and the Middle East and are virtually absent in states such as North Korea and Myanmar (Burma). Countries with authoritarian governments may pay a price for failing to unleash citizen's groups because social progress there will depend far more on the ponderous, top-down efforts of the state than on the bottom-up energy and creativity of the citizens.

The fact that governments are no longer the only important actors does not mean they are superfluous. On the contrary, governments play a critical role in preserving order, providing essential services, and regulating markets; in democracies, they provide the forum in which societies set their priorities. As the current situation in Russia illustrates only too well, when a government cannot enforce a rule of law or provide a stable economic environment, chaos reigns and most legitimate businesses struggle.

In most developing countries, only national governments (or, sometimes, foreign aid agencies) can build and maintain roads and provide other infrastructure essential to a functioning economy. The decay of rural roads in Africa and Russia, for example, is partly responsible for the continuing stagnation of agriculture in both regions. Governments also negotiate contracts for logging, mining, oil exploration, and other kinds of resource extraction—often the main assets poor countries have to sell. If they make bad deals, the country suffers both lost revenues and, usually, environmental degradation that undercuts the livelihoods of local communities. And national (and sometimes provincial) governments typically build most schools and health clinics and set educational and health policies, for better or for worse.

So it makes a difference both whether governments are competent and what policies they pursue. When they are not competent, societies suffer. Throughout Africa, ineffective governments could doom hopes for a better life for the region's people. Even in rapidly growing economies, endemic corruption can discourage foreign investment, and financial mismanagement, as in Southeast Asia, can bring economies to the brink of collapse.

Which regions will find a way to strengthen governments, to create a political consensus that can support needed actions, and to supplement the role of governments with other forms of collective civic action? The answer may be a critical factor in determining their destiny.

Nonetheless, democracy is spreading, citizen's groups are increasingly active, education and other social indicators are rising, and women are, however slowly, coming into their own. These trends, combined with economic and technological progress, suggest an optimistic future; they suggest how societies can reverse threatening environmental, equity, and security trends, initiate social and political transformations, and thus, ultimately, create a better world.

PART IV

REGIONAL CHOICES

IF THE WORLD IS BECOMING so closely interlinked that global destiny depends on regional choices, what are those choices? And if the constraints of each region—differing in history, culture, and geography—are unique, how might these factors shape each region's prospects?

Can Latin America overcome its tradition of neglect of the poor and special privilege for the rich that makes it the most inequitable region on Earth? That will require far more than economic reform.

Is it possible for China to defy its internal contradictions, to overcome its looming social and environmental problems, and to successfully complete its dash to become an industrialized market economy—and, eventually, to establish a more democratic form of government?

Can India, the world's largest democracy and likely to become the most populous nation, lift its huge masses out of their desperate poverty? That, too, will take more than economic reform.

Can the people of sub-Saharan Africa follow South Africa's lead or find other paths to effective governance and the stability needed for successful development, or will such changes come too late to avert a downward spiral of environmental degradation, malnutrition, and impoverishment—and, possibly, widespread violence and chaos?

Can North Africa and the Middle East outgrow their autocratic governments and modernize their economies and their societies, or will the struggle between rigid governments and radical Islam doom the region to continued instability and violence?

Can Russia and eastern Europe complete their perilous transition, tentatively begun, from centralized to market economies and, simultaneously, from communist states to democracies?

As for the United States, Europe, and Japan, the world's wealthiest nations and those with the fewest constraints on their futures, will they maintain open economies and provide global leadership, or will they turn inward?

The outcomes of such choices will indeed shape the world and determine whether it becomes peaceful and prosperous or polluted, impoverished, and violent.

Chapter 10

Latin America: Equitable Growth or Instability?

Stretching from Mexico and the sunny beaches of the Caribbean Sea to the towering Andes and the vast forests of the Amazon basin to the windswept Tierra del Fuego, Latin America and the Caribbean region house a growing population of nearly half a billion people. Of all the world's developing regions, this one is perhaps the most resource rich, endowed with large mineral and oil reserves, abundant sources of water, the world's largest expanse of tropical forest, and vast areas of fertile soil. Latin Americans have the longest life expectancy and the lowest child mortality of any developing region, and most Latin American countries are already urbanized, with three-quarters of the populous living in cities—about the same proportion as in western Europe.

At first glance, all the elements seem to be in place for an economic and social takeoff that could catapult Latin America into a prosperous and stable future. After a disastrous decade of debt crisis, soaring inflation, and political upheaval in the 1980s, the region appears to have changed course, with democratic governments in nearly every country and reform and modernization efforts under way. With the exception of Haiti and Guyana, no Latin American nation sits on the World Bank's list of low-income countries. All together Latin America has a $2.7 trillion economy, more than one-third that of the United States and Canada combined, and average incomes there are significantly higher than in any other developing region.[1] Brazil, which has the largest economy in the region, has a thriving industrial and manufacturing sector that is larger than South Korea's and nearly as large as China's.[2] Chile, at the forefront of economic reform in the region, has been growing at about 7 percent per year for nearly a decade and is frequently compared to the Asian tigers. Regional free-trade agreements are proliferating, and trade within the region and externally is booming: Latin America is on track to become the United States's largest trading partner within a decade.

But Latin America has stumbled before, and with its many underlying problems unsolved, a troubled future could still lie ahead. To see the fault lines that run deep through the region, one need only visit sprawling Rio de Janeiro, where slums that climb the spectacular hillsides look down on the beaches, cafés, and luxury high-rise apartments of Ipanema and other wealthy enclaves. The contrast between the elegant shops, expensive automobiles, and fashionably dressed people on downtown streets and the poverty, hopelessness, and violence of the slum neighborhoods attests to Latin America's pronounced social disparities. So does the less visible dichotomy between the often unfarmed country estates of the richest 1 percent of Brazilians, who control nearly half the country's land, and the hundreds of thousands of family farmers struggling to make a living from tiny plots or the 5 million rural families that have no land at all. Such disparities could prevent a successful future.

Despite the region's apparent prosperity, for example, widespread poverty persists. More than 100 million people, 20 percent of the population, still face the destitution and day-to-day uncertainty of absolute poverty, and many more—a staggering 40 percent of the population in a number of countries—struggle to maintain a meager existence.[3] Indeed, Latin America is by a number of measures the most inequitable place on the planet. The poorest 20 percent of its people receive only 4

Latin America

Critical Trends

A snapshot of the future based on projections of critical trends. The graphs show the plausible future range for population, economic output, and per capita GNP—a measure of prosperity. The environmental projections suggest how much polluting activities could expand, under conditions of moderate economic growth, and how many people could face potential scarcity of water and cropland, under conditions of medium population growth.

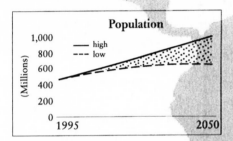

Population (Millions)
- high
- low
1995 — 2050

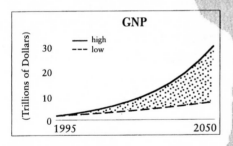

GNP (Trillions of Dollars)
- high
- low
1995 — 2050

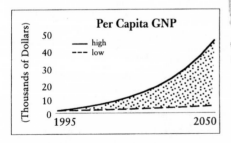

Per Capita GNP (Thousands of Dollars)
- high
- low
1995 — 2050

plausible future range

Environmental Projections

Potential growth in pollution from projected increases in energy use and industrial activity by 2050, compared to 1995 levels:

Air-polluting emissions: 250 percent
Toxic emissions: 640 percent

Number of people (in millions) facing potential scarcity of renewable resources, assuming no further degradation

	1995	2050
Water scarcity	0	60
Land scarcity	0	62

percent of the region's income, a smaller share than in any other region. Ownership of land is more highly concentrated than elsewhere. Access to education varies widely, even within countries; illiteracy in Brazil, for example, ranges from 11 percent in some urban areas to more than 55 percent in some rural areas.[4] Moreover, social unrest and rising violence suggest that stability is likely to remain elusive and support for continued economic reform could falter unless more of Latin America's people share in the benefits of progress.

Environmental problems could also darken Latin America's future. The region's megacities—Mexico City and São Paulo, both nearing 24 million inhabitants, as well as Rio de Janeiro and Buenos Aires—illustrate the challenges ahead. Already blighted by high levels of air pollution and ringed by densely packed slum settlements, these cities are ongoing environmental and social disasters. Most people at their peripheries lack such basic amenities as sewage and garbage services, making them suseptible to debilitating outbreaks of infectious disease. Children risk retarded mental development from lead pollution (most of Latin America still uses leaded gasoline).[5] But urban populations are still growing, and new megacities are expected to emerge.

In rural areas, whole ecosystems are threatened by the relentless expansion of the human enterprise as forests are cleared for cattle ranches and subsistence farms or denuded by commercial logging operations. Fires char huge chunks of the Amazon, with burning more widespread than ever before. Already, nearly a quarter of the fertile soil in Central America has been seriously degraded by erosion, much of it caused by deforestation.[6] In the Caribbean, uncontrolled development and untreated sewage are threatening beaches, reefs, and other tourist attractions that are vital to local economies. Particularly in many of the region's smaller countries, with economies still heavily dependent on agriculture, forests, and fisheries, destruction of environmental resources could undermine economic growth.

Critical Trends

How might present trends, if they continue, constrain the region's future? Demographically, although Latin America's fertility has declined, families with three or four children are still more common than families with only two children. Consequently, the region's population is still growing, and it is expected to increase by 70 percent by midcentury, to 810 million people (with a plausible range of 650 million to 1 billion). How large the population grows by 2050 will depend greatly on how successfully the region eliminates poverty and educates its people, especially its women.

If, on the one hand, the region experiences an economic takeoff and grows like an Asian tiger, then by midcentury it could conceivably become an economic powerhouse with an annual output of $30 trillion, larger than midrange projections for North America or Europe in 2050. If, on the other hand, the region returns to instability and growth falters, its economy might reach only the present size of the U.S. economy.

To see what these different economic futures could mean for the prosperity of the average Latin American requires combining economic and population projections. In the best case—with high economic growth and low population growth—average income in the year 2050 could be as high as $46,000, implying prosperity and living standards well above present U.S. levels. Such a prosperous Latin America would in all probability be a bulwark of stability; it would constitute a huge market; and the region's leading countries would very likely be major economic (and perhaps political) powers on the world stage. In the worst case, however, with slow economic growth and more rapidly expanding populations, average income might reach only $8,000— higher than at present but not enough to fulfill Latin America's aspirations. The region would be economically and politically marginal in the world of 2050, and social and political unrest might well escalate. Moreover, the effects would not be confined to Latin America: the gap between incomes in the region and those in the United States and Canada would soar far beyond today's disparity, and North America would undoubtedly face even larger migration pressure from Latin America, legal and illegal, than it does now.

Environmental conditions will deteriorate further if present trends continue, especially in the urban areas, where 85 percent of the region's population is expected to live in 2025. Even with only medium economic growth, the region's energy use—and the potential for air pollution—is expected to nearly triple over the next half century; industrial activity and the potential for toxic pollution could grow by more than a factor of six. A potential mitigating factor is that average income could reach $22,000 per person by the year 2050 under medium economic and population projections, a level that suggests the region could readily afford environmental cleanup if the public demands it.

Still, pressure on the environment will be unrelenting. Although shortages of land in Latin America stem from social causes rather than from any real scarcity, the demand for land is nonetheless likely to further exacerbate clearing of forests and farming of easily erodible terrain throughout the region. Logging activity may also accelerate, given the scarcity of timber in other parts of the world; Asian timber companies are already moving into the Amazon basin. Water scarcity will be a growing problem in parts of Mexico and El Salvador and in some Caribbean countries, such as Barbados and Haiti. Regional effects of climate change may exacerbate water shortages, with reduced rainfall a strong possibility in the Caribbean region and along the western coast of South America.

Within the constraints posed by these critical trends lie a range of possible futures for Latin America. Which of three scenarios seems most likely to emerge: a *Market World*, a *Fortress World*, or a *Transformed World*?

A Latin *Market World*

By 2010, the scope of economic reforms and the entrepreneurial energy unleashed by them created a Latin American economic miracle. Rapid, sustained growth in the region's major countries more than doubled average incomes. Although significant pockets of poverty remained, especially in rural areas, the rise in employment brought prosperity to many. Privatization of pension plans, copying Chile's successful experiment, helped raise savings, further fueling growth. Newly privatized utility companies expanded at a remarkable rate, bringing basic services—water, electric power, telephones—to many more people. A regionwide free-trade zone created such an attractive and increasingly integrated market that even the United States put its isolationist sentiments

*aside and, with Canada and Mexico, joined its southern neighbors
to form the Americas Free Trade Association (AFTA).*

*But that was just the beginning. Increased trade and enormous
flows of investment capital from the United States and from the
larger Latin nations helped to entrain even the smaller countries
in the upward trend. Rapid economic growth gave governments
the resources, and the incentives, to build new roads and other
transport links, to help turn shantytowns into permanent neighborhoods, and to buy up and redistribute idle land so as to put it
into production. Political reform and expanded sources of information—and in many countries, a decentralization of administrative authority—gave citizens more say in government decisions.*

*As the economic takeoff gathered momentum, pride in the
region's achievements and the demands of a growing middle class
brought new and largely successful efforts to control pollution, to
root out corruption, and to alleviate poverty. Education gained
priority, and governments built more schools and raised teaching
standards. A concerted regionwide effort largely eliminated the
drug cartels, forcing the remaining traffickers overseas. Birthrates
dropped rapidly. These social and environmental advances helped
to sustain rapid economic growth.*

*The Latin economic boom continued unabated for several
decades before gradually moderating. By 2050, the region was
firmly established as a bastion of democratic capitalism and a key
international player in world councils.*

Is such a future plausible? For much of the post–World War II era,
Latin America turned inward, shielding its economies behind high tariffs. Now the region is moving rapidly to join the global economy. New
free-trade zones are springing up in the Andean countries, in Central
America, and in the southern reaches of the continent. In Argentina,
Brazil, Paraguay, and Uruguay, trade has doubled every year since the
inception a few years ago, of the Mercosur (Mercado Común del Sur,
or Common Market of the South) involving those countries. Should a
hemisphere-wide free-trade zone come to pass—as some observers
think is likely—it would reinforce economic ties that are already

rapidly expanding; the Office of the United States Trade Representative estimates that trade between the United States and Latin America may surpass U.S. trade with both Japan and Europe by 2010.[7]

Attitudes have changed, too. The old "Yankee, go home!" mind-set has largely disappeared. Instead, the United States is increasingly seen as an economic model. It is the source of most of the region's investment capital, and U.S. popular culture is ubiquitous in the region.

Moreover, Latin America has some unique natural assets, including vast areas still only partially developed. Brazil, for example, is nearly as large as the United States yet has only 60 percent of the U.S. population. In the *cerrado* region of the Central Highlands of Brazil, an area larger than most of western Europe, huge grainfields march across the horizon, producing about 20 million tons of grain per year.[8] Yet only about 15 percent of the *cerrado* is now cultivated, so the potential exists to expand production by perhaps another 100 million tons per year, an amount equal to half of the world's current trade in grain. It is as if most of Iowa, Kansas, Nebraska, and Illinois were virgin prairie still awaiting the coming of the railroad—and indeed, a lack of transportation links is a major reason why the *cerrado* has not yet been exploited further. Timber plantations are only getting started in Latin America, but they may one day far exceed those that cover large areas in the southeastern United States and may export wood and paper to much of the world. In Venezuela, enormous deposits of heavy oil—still largely unexploited while the world remains awash in cheap oil—make that country potentially the regional equivalent of Kuwait.

Chile shows the potential for economic reinvention. Despite the fact that salmon do not occur naturally in the South Atlantic Ocean, Chilean fish farmers raise millions of them in pens in the cold waters off their coast; the country is now the second-largest salmon producer after Norway. In addition, Chile has exploited the fact that its summer is the Northern Hemisphere's winter, becoming a major exporter of fresh vegetables and grapes to U.S. markets.

With inflation seemingly under control, domestic consumption is rising in many Latin American countries. Multinational companies are investing heavily—building new telephone systems, expanding automobile assembly lines, adding new outlets for consumer products—often in partnership with local companies. U.S. carmakers, for example, are investing more money in the region than in China and Southeast Asia combined.[9] Mexico, despite its recent peso crisis, received more private investment in 1996 than did any developing country other than China; Brazil, Argentina, and Chile were also in the top ten.[10] Indeed, some new manufacturing efforts in northern Mexico have attained quality levels comparable to those of plants in the United States and Japan, evidence that Latin America can compete in the global economy.

Privatization of government-owned businesses is under way throughout the region as governments sell off everything from mining and telephone companies to electric utilities, seeking to stimulate greater efficiency and more vigorous growth. In fact, in a wave of social and political innovation, some Latin American countries are taking privatization and downsizing of national governments far beyond what has been done in the United States. Chile's privatized but closely regulated pension plans have been a success, increasing savings while giving people more control over their retirement funds. Higher education is even being privatized, and not just in the traditional way in which wealthy families send their children north to school in the United States. A new breed of private university is appearing, offering competition to Latin American's public universities, many of which are of notoriously poor quality.

But Latin America's successful transition to a *Market World* economy depends on several factors. Can the region grow its way out of poverty, creating enough jobs to gradually draw the poor and other marginal groups into the economic mainstream? Can the region's still fragile democracies and bloated bureaucracies evolve into paragons of efficiency and integrity? Can slimmed-down governments improve their ability to regulate private activity, protect the environment, and provide real help to the poor? If the answers to all these questions are yes, the region may experience the sustained economic boom of *Market World* and find itself in an era of widespread prosperity and stability. A close economic alliance with North America could give the two regions enormous economic power, fully matching the emerging strengths of Asia.

But markets alone almost certainly will not be enough to alleviate Latin America's widespread poverty and other social problems anytime soon or to overcome the tradition of special access to political and economic advantages for the wealthy that has helped to perpetuate economic disparities. Indeed, these disparities appear to be widening. More sweeping reforms, and more widespread political support for such reforms, are likely to be necessary; without them, the evidence suggests, there can be no economic takeoff.[11] If such a change in course—a second wave of reform—does not occur, Latin America's future may hold conflict and instability, bringing about a *Fortress World* that could cause the region to fall far short of its potential.

Fortress World

Latin America's economic boom, widely reported by the region's media, raised expectations, as did the growing advertisements for consumer goods. But as the rich and some of the middle class grew richer, the poor fell further behind. At the same time, social expenditures declined as governments downsized and corporate influence on policies grew. Urban pollution, fueled by growing numbers of automobiles and expanded industrial activity, worsened everywhere, although air conditioning spared much of the upper and middle classes from its full impact. Hardships accelerated in the countryside, too, as soil erosion made scarcity of land more acute and hunger more common. As conditions worsened for the poor, slum dwellers in the cities and landless peasants in the countryside became increasingly militant.

Global temperatures continued to climb, and the Caribbean and much of Central America suffered repeated epidemics of mosquito-borne dengue fever that caused vacationers to shun beaches and other tourist spots. Economic and social conditions deteriorated rapidly; in desperation, many island residents tried to immigrate elsewhere in the hemisphere. Protests and conflicts intensified, bringing down governments in a number of smaller countries.

The cocaine cartels and other criminal organizations expanded in scope, forging new international links, and became an ever more powerful corrupting influence on a number of the region's governments. Their effect on ordinary citizens increased, too: as kidnappings and violence became more pervasive, skilled professionals increasingly sought jobs elsewhere and transnational corporations withdrew from certain parts of the region. Respect for law enforcement and for the integrity of public officials reached new lows.

In Brazil, attempts by landless farmers to occupy unused land increased, and many turned violent when the landowners' private militias—or sometimes the local police—opened fire on the farmers. One particularly brutal incident resulted in the slaying of hundreds of people, including women and children. Captured in extraordinary detail on videotape by a journalist and played over and over again on television, the Meloras massacre, as it was called, became an international sensation. Radicalized by the videotape, many farmer's groups abandoned peaceful protest and began to arm themselves, and more confrontations occurred; landowners' villas became targets for burning. In the cities, the Meloras massacre stimulated slum communities to organize protest marches that repeatedly tied up downtown business districts and turned increasingly violent. One slum drug lord gained notoriety by offering a cash bounty for every rich person killed. Business leaders in turn suggested bulldozing the slums at the center of the protests and urged the government to restore order. The government tried, even canceling Carnaval, but the decree only prompted an even greater outpouring of rage and frustration and what increasingly seemed like open warfare between rich and poor. After more than a year of turmoil and growing paralysis, the army took power "to restore order," suspending the country's constitution.

In Mexico, a charismatic opposition politician, capitalizing on popular disgust with corruption in the ruling party and widespread economic frustration, swept into the presidency on the strength of a blatantly populist campaign. He decreed free land for all, setting off battles in the countryside, and initiated a program of massive food subsidies in the cities that destroyed the country's credit rating. Seeking to blame foreign interests for Mexico's resulting economic crisis, the president tried to nationalize all foreign-owned businesses. Removed from office by an extraordinary constitutional convention, he remained a hero to many of the country's increasingly marginalized poor.

A series of similar episodes throughout the region caused many transnational corporations to reassess—and often to abandon— their ventures. Growing tensions over illegal immigration led to U.S. withdrawal from the regional free-trade zone. Eventually, as concerns with immigration and drugs escalated and its economic interests in Latin America declined, the United States sealed its southern borders and effectively turned its back on the region. As economic stagnation set in, social conditions deteriorated further. Conflict and abrupt changes in governments added to the instability. Many Latin Americans found themselves preoccupied with personal security and even survival. The wealthy and much of the declining middle class lived behind high walls or in guarded enclaves; those who could do so sent their money, and sometimes their families, out of the region. Military governments and autocratic rulers strong enough to provide stability were openly welcomed by many.

By 2050, the tide of democracy in Latin America, once so strong, seemed to have turned. Poverty and violence were entrenched. Populations continued to rise. Once again, Latin societies lacked direction and the means to build a new social and political consensus.

In the high-rise luxury apartments of Rio de Janeiro, it is possible to think that *Fortress World* has already arrived. Here, closely guarded buildings, elaborate security systems, and bodyguards to drive businessmen to work and their children to school are a way of life even now. Wealthy inhabitants of Rio fear not just street youths who snatch purses but also armed kidnappers who target the rich. Many of the kidnappers belong to gangs headquartered in the shantytowns that ring the city. In the 1990s, Rio has been experiencing four kidnappings per week—not counting attempts foiled by bodyguards—and other kinds of violence have been increasing as well: the city's annual murder rate is 1 for every 700 residents.[12] What two decades ago was a climate of tolerance between rich and poor has broken down, replaced by fear and overt hostility. Colombia, Mexico, and a few other countries have also seen a rise in kidnapping and related criminal violence.

Potential precursors of *Fortress World* can also be found in rural areas, where scarcity, environmental degradation, and impoverishment are on the rise, sometimes leading to social upheaval. It is no surprise, for example, that growing land scarcity was a factor in the Zapatista uprising in the Chiapas region of Mexico.[13] It played a role in the long-running civil war in El Salvador, the most densely populated country in the region. In Haiti, the poorest country in the region, most of the trees in what once was a heavily forested and fertile country have been cut down, in large part because cutting wood to make charcoal has been the poor person's employment of last resort. Without trees, topsoil has washed down Haiti's steep hillsides, leaving half the country barren and unfarmable and adding to the desperation of its people and the conflict over what few resources remain. But degradation in rural areas continues, and scarcity promises to worsen. Most of the forests in Central America have been cleared, and those in southern Mexico are going fast; new logging projects are appearing in the Amazon basin.

The unequal distribution of resources such as land, characteristic of many Latin American countries, also provokes conflict. In Nicaragua, for example, land inequality was an important factor in the 1980s Sandinista uprising. In Brazil, more than half of the country's farmers have access to only 3 percent of the arable land while much of the land in large estates lies unused or is farmed unproductively—conditions that fuel a bitter sense of injustice.[14] The Brazilian constitution asserts that land should belong to those who work it, but large landholders, usually with the help of local authorities, often subvert the land-tenure process. As a result, tens of thousands of landless Brazilians, desperate for some way to make a living, have tried to settle on idle land only to be violently thrown off or killed by local police.[15] Resentment also results when those who can hire armed guards simply expropriate land and drive off those who have cleared it or begun to work it—a practice made easier because land titles are often unclear in frontier regions.

If Latin America's national governments have done little to help the poor, its state governments and legal system have done less. A lack of education among the poor, for example, contributes strongly to the region's disparate income distribution. Latin America has poor primary and secondary schools, and the region's students rank near the bottom internationally in such subjects as science and mathematics. In one survey of math and science skills among thirteen-year-olds, the only country Brazil outranked was Africa's war-torn Mozambique.[16] Attendance at secondary school averages only 50 percent for the region's population, as compared with 75 percent or higher for most of the Asian tigers. But improved equity is not the only issue: in a global economy, success depends on a skilled workforce. How can a region with such a poorly educated populace hope to compete, let alone to move more of its people into the economic mainstream?

Latin America's disparities go far beyond income or landownership or access to education. Lack of sanitary conditions makes many Latins, but especially those in the urban slums, susceptible to cholera and other waterborne diseases. The region still has 10 million malnourished children, mostly in Brazil and Mexico.[17] The poor bear the brunt of air pollution and exposure to toxic chemicals—which are on a rising trend.

Despite the rise of democracy, most governments in Latin America have neither shifted priorities to redress such disparities nor proved effective in social reform. All too often, social programs in the region do not target the poor; most subsidies benefit others. Chile is a notable exception—in three years, it reduced the number of people below the poverty line by one-quarter.[18] At the same time that social conditions fester, military budgets remain high—despite an absence of real threats to countries in the region—reflecting an ongoing military influence. Costa Rica is the standout exception here: for more than three decades, it has had no military force of any kind, just a police force, and in its economic and social development it ranks far ahead of its Central American neighbors. Inefficient and often corrupt tax collection denies many governments large additional revenues.

Not surprisingly, tolerance for growing disparities may be wearing thin in many parts of Latin America, and disillusionment with economic reforms is growing.[19] An independent press and, especially, television make such disparities and the conflicts they engender far more visible. Political support for economic reforms, let alone more difficult social reforms, could easily fade unless these reforms produce widely shared results, which are not yet in sight. Only in Chile and perhaps Costa Rica is the process far enough along, with widespread support for reform efforts, to render political reversals unlikely. The price of even temporary reverses or stumbles can be severe. The 1994 peso crisis in Mexico and the more recent financial crisis in Southeast Asia have shown all too clearly that governments now face high expectations from international investors, as well as internally, yet have less maneuvering room in a global economy.

If *Market World* in Latin America fails—if the current wave of economic reforms does not generate sustained growth, if social investment and government reform remain inadequate to bring the poor into the economy, if powerful criminal enterprises corrupt whole societies—then the likelihood of more conflict, violence, and instability will rise. In that event, it is entirely plausible that outside investment will dry up, that the rich will retreat into secure communities and landed estates, and that history will repeat itself with armed insurgencies, populist governments, and military takeovers, undermining economic reforms and democratic governments.

Is there another alternative?

Latin America Transformed

The wave of economic reform that triggered the start of Latin America's boom times was striking. Yet even more remarkable and unexpected was the second wave of reform, which blurred Latin America's sharp divisions between rich and poor.

The transformation was both top-down and bottom-up, driven in some cases by visionary leaders and in others by community groups organized or supported by the church, nongovernmental organizations, or governments. In Peru, for example, President Alberto Fujimori's most enduring legacy to his country turned out to be the massive program of school construction and other public works targeting the poor that he pushed through in his third presidential term. Together with new laws giving clear land titles to rural farmers and urban slum dwellers, this effort opened radical new opportunities for the lower classes in that country. In Brazil, a similarly visionary president brokered a historic deal between the landed elite and landless farmers, transferring more than a third of the nation's prime but underused or unused farmland into eager new hands and shares of newly privatized industrial companies into the hands of former landowners.

Bolivia's radical decentralization of political and economic power, giving communities a major share of national tax revenues and the power to shape local development, seemed a dubious experiment at first. But eventually, the reforms galvanized communities and small towns into action and spurred entrepreneurial activity, generating thousands of new ventures that are now described as the Latin equivalent of China's township enterprises. Prosperity rose rapidly, especially among the poorest segments of Bolivian society. Many Central American countries adopted similar strategies, which stimulated a remarkable economic and social renaissance. The region's common market and its bold decision to follow Costa Rica's lead and disband national armed forces aided the process.

Accompanying Latin America's political and social transformation has been a dramatic shift in attitudes and values. Central America, the most environmentally devastated part of the region, has embraced a new sense of stewardship of nature; ecotourism now drives a major part of its economy. Mexico's passage of anticorruption laws that allow law enforcement agencies to examine all private financial transactions by elected public officials and judges has set a new standard of public accountability that is widely admired and copied. Tough new antidrug efforts, supported by citizen's groups, appear to have the cocaine trade on the run.

Latin America's political and social transformation did not happen overnight. Even in 2050, some poverty remains. But educational levels and other social indicators are greatly improved, and with them economic mobility. It is now clear that social programs and other changes did not hold back economic progress, as some businesspeople had feared, but instead extended and accel-

erated the economic boom. The highly skilled Latin workforce is fiercely competitive, and those who are no longer poor buy a lot of goods, making Latin America a consumer market to rival China. Even more important, the region's political and social stability is no longer in question. In many ways, Latin America has become a model for how a multiracial society can succeed.

Such a *Transformed World* will not come easily. Achieving it would mean putting aside the region's historical neglect of the poor and creating a more equitable society. It would mean overcoming the opposition of the wealthiest and most influential segments of society to break up unproductive landholdings. It would mean changing spending priorities and establishing more focused efforts to alleviate poverty and provide basic services. Ultimately, it would mean making profound changes in the basic political structure of Latin societies. But there are precedents for such measures. Drastic land reform helped initiate decades-long economic booms in South Korea and Taiwan.[20] Moreover, the ongoing privatization of government-owned companies means that Latin American governments could offer their big landowners a swap of landownership for stock in newly privatized companies rather than just forcing them to sell or abandon their land. Such a move might invigorate both rural land markets and stock markets. Chile, as described earlier, has already shown that targeting social programs to the poor can rapidly have profound effects. Bolivia is in the middle of a revolutionary effort to decentralize government, a sweeping transfer of power from the national government to small towns and rural communities that, if it succeeds, will invigorate local economies and grassroots democracy.[21] Costa Rica has already made environmental protection a priority, setting aside and protecting large areas as ecological reserves—what the country calls "factories of biodiversity" for the future. Throughout the region, community groups and other nongovernmental organizations are becoming increasingly effective and numerous. There are dozens of national environmental and social development groups that frequently collaborate, as well as thousands of grassroots community groups, providing an organizing vehicle and a voice for many who have traditionally been excluded from Latin American decision-making processes.

Yet as promising as these developments are, Latin America lacks many social and political reforms critical to a *Transformed World*. Still needed are clear titles to land in informal settlements and frontier farms, since property rights are basic to land markets and offer incentives to improve, rather than degrade, land. Lacking, too, is a broader respect for human rights and justice in the form of impartial courts and a fairer and more accessible judicial system. Pollution laws are not yet routinely and evenhandedly enforced. As Alvaro Vargas Llosa, a best-selling Latin American author and journalist, has observed, "There is a grave danger in the fact that the courageous but limited experiments in reform . . . are giving rise to powerful populist oppositions and weakening the case for the free market precisely because [these reforms] are so incomplete."[22]

Is a *Transformed World* scenario plausible in Latin America? Perhaps. The political will and wider shifts in values necessary to bring it about are not yet widespread, but the beginnings are there. It depends on the choices Latin America's people make in the years to come.

Chapter 11

China and Southeast Asia: Can the Asian Miracle Continue?

FROM THE FRIGID NORTHERN REACHES and vast interior deserts of China to its booming coastal provinces, from the dense forests and teeming waters of the Mekong River basin to the immense archipelago of Indonesia, Eastern Asia and Southeast Asia house nearly a third of the world's population. The destiny of the world is thus inseparable from what happens in this part of Asia: demography alone ensures that. But the economic and political stakes are also high. Despite current difficulties in a number of countries, these two regions are home to the fastest-growing economies in the world, whose combined output will almost certainly overshadow even the U.S. economy in coming decades. The development strategies followed here are closely watched by other regions: their success or failure will have widespread repercussions. Will the regions' largely authoritarian governments, for example, eventually yield power and become democracies, or will they continue to hold out to the world an "Asian model" of governance—of societies in which political freedoms remain tightly constrained or nonexistent?

There is no denying that both these regions have markedly raised the prosperity of their populations. Not long ago, the original Asian tigers—South Korea, Taiwan, Singapore, and Hong Kong—were poor countries with average incomes lower than those in many African nations. Now, after nearly four decades of rapid growth, these countries have average incomes comparable to those of some European countries. A second wave of would-be tigers, including Thailand, Malaysia, and Indonesia, has also fared well—despite the current downturn—and the Philippines and Vietnam are beginning what could be a third wave of growth in the region.

China's record over the past two decades is even more spectacular. The Chinese economy has grown more than twice as fast as that of Japan and more than three times as fast as that of the United States. Already, China is shaping corporate strategies and affecting consumer buying patterns around the world. Most athletic shoes, regardless of brand, bear the imprint "Made in China," and some forecasts suggest that China's economy could become the largest in the world as early as 2020.[1]

Yet the future of China and Southeast Asia is not assured. Both regions face growing social and environmental problems brought about by their rapid economic expansion—problems that could undermine their success. Corruption is rampant, and the gap between the wealthiest beneficiaries of the boom, mostly in the cities, and the vast numbers of rural poor is widening. Moreover, most of the regimes in these two regions have staked their political credibility on their booming economies, in effect promising prosperity in return for nearly uncontested power—a dangerous social contract that, if it unravels, could lead to instability. Managing growth is thus the critical challenge for both of these regions as well as the reason that, despite their many differences, I discuss them together.

With cultures that encourage hard work, thrift, and respect for family, and governments that for the most part have a track record of pragmatism and competence, both China and Southeast Asia have bright prospects. But with their volatile histories, smooth and uneventful

China

Critical Trends

A snapshot of the future based on projections of critical trends. The graphs show the plausible future range for population, economic output, and per capita GNP—a measure of prosperity. The environmental projections suggest how much polluting activities could expand, under conditions of moderate economic growth, and how many people could face potential scarcity of water and cropland, under conditions of medium population growth.

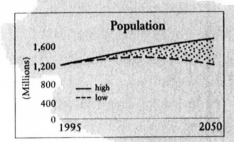

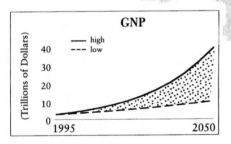

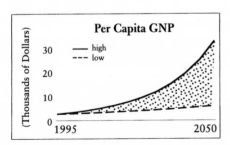

☐ plausible future range

Environmental Projections

Potential growth in pollution from projected increases in energy use and industrial activity by 2050, compared to 1995 levels:

Air-polluting emissions: 600 percent
Toxic emissions: 1060 percent

Number of people (in millions) facing potential scarcity of renewable resources, assuming no further degradation

	1995	2050
Water scarcity	—	—
Land scarcity	0	1517

growth might be the most surprising scenario of all; indeed, the 1997–1998 financial crisis in Southeast Asia underscores the regions' vulnerability. Moreover, political instability or conflict within either region would have international implications—China is a nuclear power with more than 2 million soldiers under arms; North and South Korea, Vietnam, and some other countries of Southeast Asia also have large armed forces.

For both of these regions, a number of different futures are plausible. But the most interesting, most uncertain—and globally the most consequential—future is that of China.

China: The Giant of Asia

Shanghai, like many of the rapidly industrializing and modernizing parts of China, seems on first inspection to be one vast construction site. According to a recent estimate, greater Shanghai has one-fifth of the world's high-rise construction cranes now in operation. Even if that number is slightly exaggerated, the scale of construction here is nothing short of astonishing. Among the projects underway in 1997 were a second large international airport; a new deepwater port larger than that of Hong Kong; new highways, bridges, subways, and rail facilities; and many square miles of factories, modern shopping malls, high-rise apartment buildings, and high-tech office complexes—including what will become (at least temporarily) the world's tallest building. Rough-clothed migrants from China's interior, pouring into the city to work and to gawk at stores laden with Western consumer products, mingle in the streets with sleek-suited foreign businesspeople and fashionably dressed urban Chinese. For a city that fifteen years ago had a rather old-fashioned look and was virtually closed to outside commerce, the transformation is stunning. Early in the new century, Shanghai will be a world-class city, an apt symbol of China's ambitions.

China's abrupt turn toward a market economy has meant an equally abrupt transition for many of its people. One example is found in the story of Chen Jinhai, who began life as the eldest child of a landless peasant in Pengdong, one of the poorest villages in Wujiang County, west of Shanghai.[2] At age six, he rose every morning at four o'clock to help catch crabs, the family's main source of income; he dropped out of school after two years of middle school. But Chen was not content with peasant life. In his twenties, just as China began its move to a market system, he apprenticed himself in a fiberglass factory, sleeping in a public bathhouse because he had no money. Overcoming many obstacles, he founded a two-room village factory of his own, which has grown into the Wujiang Fiberglass Group, with seven factories and more than 350 employees. The company was one of the major subcontractors in building Shanghai's subway system and now exports its products to other Asian countries. Chen, now in his forties, has become a wealthy man, a multimillionaire.

Rags-to-riches stories such as Chen's are common among the entrepreneurs and small-scale industrialists who have fueled China's almost miraculous economic growth.[3] But that growth has also improved the welfare of many of China's people, lifting more than 200 million people out of poverty in just twenty years. Following a strategy of export-led growth similar to that which made Japan an economic superpower, China is already a major exporter of shoes, textiles, toys, and other manufactured goods—and, like Japan, it is building up huge trade imbalances with the rest of the world. But China is also building up its industrial might to supply its own needs. It is now the world's largest steel producer and will soon be the world's largest chemical producer. Endowed with abundant coal reserves, China aspires to build fifty new coal-fired electric power plants per year for the foreseeable future, along with other massive energy, transport, and telecommunications projects. If its growth continues, China will have modernized its economy and transformed the lifestyles of its people within a couple of generations—an impressive feat.

China's achievements are not just economic. The country has dramatically lowered its fertility rate to 1.9 children per woman, below the replacement rate needed to eventually stabilize its population. At the same time, agricultural production has soared, rising by more than 50 percent in the 1980s; grain yields are approaching those of the United States. As a result, the specter of famine, which once loomed large, is now fading. Equally remarkable are China's rapidly rising literacy rates, high level of school enrollment, and generally high level of health care.

Democratic elections to choose local leaders are now standard practice in the nearly 1 million villages within China, even though the provincial and central governments remain under Communist control. That China, still a poor country by any standard, has accomplished so much so quickly is a source of hope for similar transformations in other developing regions.

Perhaps most striking is that China's rapid development has occurred amid internal tensions and seeming contradictions. The country has the world's most rapidly developing market economy, yet it is ruled in an authoritarian manner by a Communist Party that harshly represses all dissent. Its new wealth is generating huge disparities between urban and rural areas, between coastal and interior provinces, between the newly rich business class and average Chinese workers with an income of a few dollars a day—disparities that contrast sharply with what had been an ideology of equality and a spirit of shared sacrifice. And China, with its harshly imposed but relatively successful "one-child" policy, is the only country ever to undergo a demographic transition (to lower fertility below the replacement rate) while its population is still overwhelmingly rural and relatively poor.

Although China's outlook is largely optimistic, it is not unclouded. For example, other than Japan, China is the most resource constrained of any of the world's major nations—it feeds 20 percent of the world's population with 7 percent of the world's cropland; it faces critical shortages of water, especially in its northern and western provinces; and it has only 4 percent of the world's forested area. Industrial pollution has expanded with the economy; air pollution is already severe in most large cities, and many of China's rivers carry so much toxic waste as to pose a severe public health hazard. Yet shutting down the most polluting factories, many of them too small to install cleaner technology, risks throwing millions of people out of work. So does privatizing or closing China's money-losing state-owned industries, of which there are some 300,000, including most of the country's heavy industries.

Poverty, although much reduced, has not disappeared; the World Bank estimates that 250 million Chinese still make less than $1 per day, with 70 million living in absolute poverty. The banking system is nearly bankrupt, a consequence of forced loans to state-owned enterprises.

The flood of urban migrants seeking jobs—80 million people by the government's estimate—has jammed transport facilities and overwhelmed cities; in the 1990s, huge urban slums appeared. Economic and legal reforms required for a full-fledged market economy are far from complete and some, such as private ownership of land, pose ideological hurdles too high for the current government to leap. And although some decentralization of authority has occurred, giving provincial authorities greater independence from the central government, democratization and other political reforms are not yet in sight— the Communist Party tolerates no political opposition.

Critical Trends

How might present trends, if they continue, constrain China's future? China's population, at 1.2 billion now the largest of any country in the world, is not expected to grow much more and could stabilize at about 1.5 billion as early as 2025; by 2050, the population could conceivably decline again to its present size (the plausible range is 1.2 billion to 1.765 billion). Thus, if present trends continue, China faces relatively less population growth than any other developing region. But it faces another kind of demographic pressure, with an estimated 300 million urban migrants expected to crowd into cities by 2010; even more will follow as China shifts rapidly from an agricultural, largely rural country to an industrialized, largely urban one.

How prosperous could China become? Even if China's torrid pace of economic growth slows significantly, its economy might still increase nearly sevenfold by the middle of the next century (the medium growth projection), and could match the U.S. economy in size. Under optimistic, high-growth assumptions, the economy could grow elevenfold, becoming the largest in the world. China's growth could also falter.

In the best case (with high economic growth and no net population increase), average annual incomes in China would reach $33,000 by the year 2050. Although this figure represents real prosperity, it is less than projected levels for either Latin America or Southeast Asia under their best-case assumptions. Nonetheless, such a prosperous China would almost certainly be a major world power. Under more moderate, midrange projections, incomes would reach nearly $16,000 per year—

perhaps enough to bring the bulk of the country's population a middle-class lifestyle. And in the worst case, as might result from an extended period of economic and social chaos, average incomes could be as low as $6,000.

Environmental conditions in China could become intolerable if present trends continue. Urban air quality is already unhealthy, with pollution levels well exceeding the World Health Organization's guidelines in many of China's cities, largely because of the country's reliance on coal. But even midrange economic growth is expected to require a sixfold increase in energy consumption over the next half century, and high growth would mean even more.[4] If expanded energy consumption were to translate into a comparable increase in urban air pollution, many cities would be literally unlivable, yet China has few alternatives to coal.[5] The effects of China's rising use of coal would also extend far beyond the country's borders—by 2020, China is likely to become the world's largest source of potentially climate-altering greenhouse gases. Uncontrolled dumping of toxic chemicals is already an important environmental issue in China, creating significant health problems, and the country's industrial sector—and thus, potentially, its output of toxic materials—is projected to increase more than tenfold over the next half century. So China has no choice but to use cleaner technologies and enforce antipollution laws, as the country's leaders recognize, and to do so far sooner in its development and at far lower levels of average prosperity than have other rapidly industrializing countries. Even so, pollution is likely to get worse before it gets better.

China also faces the prospect of increasingly scarce farmland and water. A new U.N. study, for example, suggests that by 2025 the country's rising demand for water will run into serious restrictions. Already, water clean enough for drinking or even for industrial purposes is in short supply in many parts of the country. And climate change may exacerbate China's water problems, reducing rainfall in the already arid interior portions of the country and increasing the risk of flooding in other areas.

Social and political factors may also constrain China's future. City dwellers now earn, on average, three times as much as rural Chinese, and their incomes are rising twice as fast. Moreover, the lifetime guarantees that once accompanied jobs in state-owned industries are disappearing, with the result that some urban residents find themselves begging on the streets for as little as $.50 per day—often not far from Beijing's private clubs that, with annual membership fees of $10,000 or more, cater to China's newly wealthy.[6] Such disparities between the newly rich and those still poor threaten the country's sense of shared sacrifice and could eventually undermine its political stability. At the same time, growing economic freedom and wealth are likely to fuel demands for political liberalization. Successfully managing China's growth under such constraints will require extraordinary skill on the part of China's government.

As these trends suggest, a range of very different futures is plausible for China. Which one will prevail—a *Market World*, a *Fortress World*, or a *Transformed World*—and what might be the consequences, for China and for the world?

Market World

The economic reforms begun in the old century continued at an even more rapid pace in the new. Over a decade, virtually all state-owned enterprises were closed down or transferred to private hands and the banking system was modernized. New tax systems allowed urban areas to build housing, roads, and sewage systems to cope with the continuing influx of migrants, with the construction boom absorbing many of the laid-off industrial workers. Farmers were allowed to own the land they farmed, prompting many of them to sell it off and hastening a consolidation and modernization of agriculture. And foreign investors continued to pour money into new factories. As a result, China's headlong economic expansion—and its transition to an industrial market economy—continued.

As prosperity spread, poverty largely disappeared. But the demands of a rising middle class for improved social services and less-polluted cities, of foreign firms for a more consistent rule of law, and of China's trading partners for more open markets began to make themselves felt. Pressure for political reform, for more responsive government, began to grow.

Continuing tensions and disputes between provincial authorities and the central government led to the adoption by the Party Congress in 2013 of a remarkable document that, for the first time, defined and limited the powers of the central government—creating what was, in effect, a constitution. Subsequent Congresses added a bill of rights. Claiming its powers under the new document—and implicitly, its need for parity with Hong Kong to compete effectively—the Shanghai region created a provincial legislature and, five years later, began direct election of provincial government officials. Other provinces gradually followed, and by 2040 China was a democracy.

By the middle of the century, China was a respected voice and often a leader in global affairs, a powerful economic competitor, and a model for many developing regions. It had become the world's largest market for and producer of automobiles and many other consumer products, a world-class manufacturer with a global reputation for quality and innovation. Increasingly, Chinese was becoming, along with English, the global language of business, and commentators were beginning to talk about the Chinese century.

Most economic analysts expect China's rapid economic growth to continue, even if at slower rates, and forecast that China will become wealthier and more powerful than it is today at the end of the twentieth century. Some analysts also argue that, along with economic liberalization and the rise of a middle class, political liberalization is inevitable, too. The conventional wisdom for China is, in effect, a *Market World* scenario. There is no need to consider this hopeful scenario in more detail—one need only read nearly any article on China in a business magazine or the business section of a newspaper.

But economic and political reforms cannot be taken for granted; nor will a *Market World* future necessarily solve China's environmental, social, and political problems. What might compromise China's future? Two possibilities seem most likely: urban unrest (compounded by massive pollution and widespread unemployment) and military conflict with its neighbors (propelled, in part, by growing resource scarcities).

As an introduction to the first scenario, I begin with the story of

Chunmei Yuan, who was seventeen when she talked to Zhihong Xiong, a Chinese journalist who helped me find examples of that country's urban explosion. A year earlier, Chunmei, the eldest daughter of a family in Henan Province in central China, had faced a life of farm work in her small village. But Chunmei wanted a life more like what she had seen in movies and on television, and she resolved to go to Beijing to seek work.

It turned out to be easy for Chunmei to find a job in the city because many urban residents need maids and baby-sitters and family service companies have sprung up to supply them. Barely a day after Chunmei arrived and registered with such a company, a family hired her as a live-in baby-sitter; after six months, satisfied with her work, they renewed her contract and gave her a raise.

Chunmei finds her new life far more interesting than life in her village, she has better food, and she hopes to save enough to buy some fancy clothes and a new bicycle. It's not that everything is perfect: a vendor who sized her up as a rural girl robbed her, and she didn't dare complain or tell anyone; sometimes she has been scared almost to death. But she is planning to write to her younger sisters and tell them, "No matter what, try to find opportunities to go to cities to see the world outside the village, so you won't live life like a fool."

Many young people are taking Chunmei's advice these days and seeking their fortunes in the cities, even when their education prepares them for little more than child-care work or construction labor.

Fortress World I: The Revolt of the Urban Masses

> *The 1990s were merely a prologue. China's massive urban migrations continued well into the new century as its people sought to escape the hard, limited lifestyle of rural areas—a lifestyle made harder by growing pollution and environmental degradation—and to find higher wages and greater opportunities.*
>
> *But although urban wages were higher, most migrants did not have the education or skills for more than menial jobs, nor did the overburdened cities have a place for them to live. The migrants created huge shantytowns that soon became dangerous, crime-*

ridden settlements often beyond the control of authorities. With no water or sewage services, the shantytowns became a source of pollution and disease, and under China's archaic laws, cities saw no obligation to offer their inhabitants social services.

But as long as urban economies could provide jobs for the newcomers, the urban masses were for the most part politically docile; despite an often miserable lifestyle, they had few thoughts of returning to the rural villages from which they had come. As the growth of the Chinese economy slowed after 2010, however, and economic downturns became more frequent and more severe, urban unemployment and unrest grew. The unrest was exacerbated by increasingly dreadful environmental conditions: polluted water supplies, growing toxic emissions from China's now massive industrial base, and often life-threatening air pollution. The cities were blanketed with clouds of smog and dust so thick that the sun often could not be seen for weeks on end, so that everyone who could afford to do so wore filter masks.

The riots of 2015 began in greater Tianjin when residents of the urban shantytowns marched into the city proper demanding food, cleaner air and water, and jobs. Confronted by police and eventually the army, the rioters dispersed throughout the city, looting stores and even homes as they went. Attempts to burn the shantytowns led to pitched battles and many casualties, but not to peace. Efforts to supply free food and to buy off the slum dwellers seemed to make matters worse. Indeed, as the riots rapidly spread to other regions, the rioters began to demand political reforms as well, toppling a number of city and regional governments. The central government intervened by mobilizing the army and promising improved conditions, which quieted the riots that year. Clearly, however, a new and volatile political force had entered Chinese life.

Economic hard times continued, and the riots came again; conditions were simply too wretched and the slum dwellers too numerous for the government to meet all their needs. As the slum dwellers' movement grew, wall posters began to call for radical change, for a true workers' government, and to quote Marx.

Faced with widespread social unrest and violence, foreign invest-
ment stopped and many foreign companies fled to quieter regions.
China's future, once bright, looked increasingly bleak.

It may seem a remarkable irony to bring up the possibility of a revolt by an urban proletariat in a nominally Communist country, but the possibility worries both Chinese authorities and foreign analysts. The massive migration of people to China's cities is real, involving some 50 million people between 1990 and 1995, a projected 100 million more by the year 2000, and an additional 170 million by 2010.[7] So are China's growing pollution problems.

The Huai River, for example, rises in the mountains of central China's Henan Province and flows for 700 miles, through 182 of China's most densely populated counties, before it discharges into the East China Sea. Along its banks and those of its tributaries live 150 million people, many of whom depend on the Huai for drinking water and other household uses, for water to irrigate their crops, and, increasingly, for water needed by the thousands of township industrial enterprises that have sprung up in recent years. One of these, the Tongbai Paper and Pulp Mill in the sleepy town of Tongbai near the Huai's source, illustrates what has happened to the river and to those who live along it.[8]

The paper mill provided employment for hundreds of Chinese, and in 1993 it helped to boost Tongbai's industrial output to $20 million. But in that same year, it also discharged more than 2.3 million gallons of untreated wastewater into the river. Nor was it alone. According to official reports, more than 7 million tons of pollutants are discharged into the Huai every day, more than 2.5 billion tons per year. Much of the river has become so black and foul that it is literally unusable for most purposes; in 1994, there were four separate pollution emergencies requiring a cutoff of tap water, closure of factories, and trucking in of water from other regions for human consumption.

The incidence of a wide range of health problems along the Huai and its tributaries, from liver and spleen problems to cancer and birth defects, is far higher than China's national average. The river's fishery, once very productive, is now effectively destroyed, and crops and plants no longer grow on land near many parts of the river; poisoned livestock are common.

The Huai is far from an isolated case. Toxic emissions from outdated factories, combined with runoff of pesticides and fertilizers from farmlands, have poured into the country's rivers. Indeed, 78 percent of China's waterways are so badly polluted, according to the government's 1996 *State of the Environment* report, that their water is not suitable even for industrial purposes. In addition, an estimated 40 million Chinese fall ill every year from contaminated drinking water. Belatedly, the government is making an effort to clean up the Huai, closing thousands of paper mills and, according to government sources, reducing the volume of pollutants flowing into the river by about 10 percent. Nonetheless, new factories continue to sprout, and government policy, especially at the province level, still tilts heavily toward development.

Polluted water is hardly China's only environmental problem. Widespread use of coal as a fuel has created acid rain that is damaging crops and forests. In urban areas, tenacious air pollution, also largely from the combustion of coal, has made chronic respiratory ailments common. A World Bank study estimates that as many as 80,000 deaths will be caused by air pollution in Beijing alone by 2020 if current trends persist.

Such rampant pollution may be the consequence of a new and still raw market economy outrunning the ability of the Chinese government to create and enforce environmental laws. But it also reflects China's preoccupation with economic growth, captured in Deng Xiaoping's admonition that "getting rich is glorious." Will improvements in regulation and more modern technology be enough to offset China's rapidly expanding industrial economy so that pollution levels recede rather than worsen? That remains to be seen. But clearly, the environmental crisis in China is having a devastating effect on human health and welfare. Moreover, since the health effects of exposure to toxic chemicals often show up years or even decades later, China may be creating a horrific toxic legacy, a Chinese "silent spring," for its people.

At the moment, such concerns do not seem to weigh heavily on most of China's people, or, indeed, on their leaders, but public opinion could change rapidly. In the Soviet Union, concern over environmental and related health problems—brought to a head by the Chernobyl nuclear reactor accident and official ineptness in dealing with it—played an important role in undermining the credibility of the Soviet government and contributing to its fall.

Internal problems could thus derail the Chinese economic miracle. But external problems could, too, if China were to become a security threat to its neighbors.

Fortress World II: The Siberian Excursion

In the decade following Deng's death, China's aging leadership became increasingly concerned with what it perceived as a new set of threats to the country's security: a growing dependence on imported oil, timber, and food, traced to the country's limited stock of fertile land and other natural resources; and the threat to social stability posed by the growing flood of internal migrants from its impoverished and water-short interior and other rural areas.

Shortly after the Party Congress in 2005, China began to "encourage" massive emigration to Siberia. Many Chinese, employees of the Japanese and U.S. companies harvesting the region's natural resources, were already in Siberia, but the new policy increased the flow and helped them convert temporary work camps into permanent settlements. Chinese companies also bought a share of the ownership in many of the resource extraction activities and diverted an increasing share of their output to China. There was sporadic Russian opposition, official and unofficial, to the new settlers, and before long, China also sent troops (designated as "local militia") to protect the settlers. Within a decade, despite occasional violence and very harsh conditions, more than 1,000 Chinese settlements with close to 25 million people were scattered across southeastern Siberia; they increasingly dominated the long peninsula above Vladivostok and reached almost as far north as Yakutsk.

The Russian government, preoccupied with its own economic and political struggles, initially did little about the Chinese move into Siberia, especially as the Chinese government was careful not to challenge Russia's sovereignty over the area or disrupt its access to its Pacific ports. But by 2015, the flow of timber, minerals, and oil southward into China over newly built rail lines and pipelines had become too large to ignore, as had China's increasingly direct administrative control over the area. Charges by Russian ultranationalists that the moderates had "lost Siberia" led to formal demands for Chinese withdrawal and threats of expulsion by force. The Chinese responded by moving mainline army units and weapons into Siberia and halting rail traffic to Vladivostok.

After months of skirmishing and tension, the weakened Russian military persuaded its government that it could not project enough force into the region and might fail if it tried to drive out the invaders. A standstill ensued, with China in de facto control of a large portion of eastern Siberia. The international community had no stomach for a military response and, despite a halfhearted attempt at a trade embargo, could not put together enough of a consensus to impose effective economic sanctions on China. Nonetheless, foreign investment in China plunged precipitously; new factories and other facilities were postponed. Trade suffered, and although it did not cease entirely, China was increasingly forced to rely on its own limited resources.

Meanwhile, the Russians launched a major effort to rebuild their military forces, and in 2022 Russia surprised China with missile and air attacks that destroyed dozens of the larger Siberian settlements—killing tens of thousands of civilians—and severed virtually all rail, road, and pipeline connections between Siberia and China. The situation threatened to develop into full-scale (and possibly nuclear) war between the two countries. Frantic international mediation efforts averted that apocalypse, but for years the situation remained tense and, on the ground, violent as the Russians retook large portions of the Chinese occupation zone.

The prolonged crisis brought China's economic growth to a standstill. By 2030, the country was producing little more than in 2010, and living conditions had deteriorated. Dissent was fiercely suppressed, however. Internationally isolated, China effectively sealed itself off and withdrew from the world community.

There are already some Chinese workers in Siberia, but there is no evidence that China's *current* leadership would undertake such a Siberian excursion. Nonetheless, there is ample potential for conflict between China and the rest of the world over trade, over human rights, or over natural resources. For example, China has staked claims to potential oil fields in the South China Sea that are also claimed by Vietnam and the Philippines. And, perhaps most dangerous, China maintains (and insists that other nations also maintain) the diplomatic fiction that Taiwan is a province of the mainland. China has repeatedly threatened to invade Taiwan if the country ever declares its independence, so if a Taiwanese government does so, the United States might face the prospect of either engaging in armed conflict with China or abandoning Taiwan. So long as China has a totalitarian government, it represents a potential security risk to its neighbors—and a potential threat to its own future prosperity.

Moreover, the resource constraints alluded to in the scenario are real. How China deals with them will inevitably affect not just its future but also that of the world. Cropland is already scarce, at 0.08 hectare per person, or less than 1 acre for every five people, about the same as in Bangladesh, and is expected to be only 0.06 hectare per person by the year 2050. Yet China is expected to add several hundred million more mouths before population growth stops, and rapid urbanization and industrialization are converting prime farmland at a rate that could reduce the amount available by 10 percent within another two decades. The high population growth projection—plausible if the central government's control over its citizens gradually eases and Chinese couples use their new freedom to have more than the one child now officially allowed—would only make things worse.

Water—polluted or not—is scarce in the northern and interior regions; according to Vaclav Smil, a scholar at the University of Manitoba, 50 million people already do not have enough water for basic daily use, and irrigation shortages affect an area larger than France. Yet most of China's vast coal deposits lie in the same region, so plans to triple coal production over the next two decades will further worsen water shortages.

If China adopts the Japanese strategy—to solve cropland and water scarcities by importing much of its food (and perhaps other resources, such as wood and oil) and to pay for those imports with manufactured goods—the global impact could be enormous. Lester Brown, in his book *Who Will Feed China?*, predicts that China may import as much as 400 million tons of grain (as opposed to about 6 million tons now), more than world food markets can easily supply.[9] Importing such enormous quantities would drive up world food prices and, in effect, deny food to poorer regions such as Africa that are already dependent on imports and food aid. Most agricultural experts, however, question Brown's figures; their consensus is that China will remain largely self-sufficient.

But if food is not limiting, what about the effect of massive imports of energy and other natural resources? China has very meager oil reserves. If it builds a road-based transportation infrastructure and, eventually, a fleet of a few hundred million vehicles—still far below U.S. and Japanese levels on a per capita basis—then its oil imports and those of the rest of Asia could strain global capacity. A China that imports as much wood as Japan now does, even though it would represent only one-twelfth the Japanese per capita consumption of wood and paper, would create enormous pressure to log forests in Asia and Africa that are already rapidly disappearing. What is inescapable about the future is that China will have a huge environmental impact, not only on itself but also on the world as a whole.

Still, these are not insuperable problems. Japan managed to become an economic superpower with fewer natural resources than China. If China can manage its transition to a market economy, including the social and political reforms that are likely to be necessary, could it solve its resource scarcities and some of its environmental problems by adopting—even helping to develop—advanced technologies?

Transformed World: Chinese Technological Leapfrogging

The solar-cell program was propelled in part by opposition to the construction of more coal-fired power plants near urban areas already choking on the particulates and sulfur-laden gases they emitted. The Chinese government, noting the success of U.S. and

Japanese research and development efforts to improve cell effi-
ciencies and bring down manufacturing costs, targeted a major
portion of its R & D money for that technology. Within a few
years, dozens of research institutes were devoted to duplicating
and extending the foreign results and setting up small manufac-
turing sites.

When the Arab-led oil cartel raised oil prices sharply in 2010,
however, the program became a strategic priority. The Chinese
government offered leading solar-cell manufacturers partial subsi-
dies and a guaranteed market if they would build advanced
demonstration plants in partnership with Chinese firms. Three
firms—one French, one U.S., and one Japanese—accepted, and
China became the world's largest market for solar cells.

In 2018, after nearly a decade of experimenting with ever
more efficient cells—now mostly of Chinese design—and
advanced manufacturing processes, China opened a full-scale
industrial factory of its own. The heavily automated facility
dwarfed any other solar-cell factory in the world and was compa-
rable in size to the largest semiconductor plant then in existence.
A second, still larger factory opened two years later. By 2025, the
country had more than a dozen manufacturing sites turning out
many square miles of cells each year—the equivalent in generating
capacity of fifty major new power plants annually—with more
manufacturing sites opening every year.

Installation of solar panels, officially encouraged and at first
partially subsidized, took place on rooftops of private homes, on
municipal and industrial buildings nearly everywhere, and in giant
arrays adjacent to new industrial parks. Installation, maintenance,
and repair of solar panels became a vast area of employment, and
becoming part of a "solar cadre" was a popular aspiration among
Chinese youth. Combustion of coal, and pollution levels, began to
decline.

The solar energy boom helped fuel China's growing industri-
alization, especially in the country's interior. Its vast, arid deserts
became a valuable resource, providing the sand to make solar cells
and the sunlight to use them; the deserts began to export energy

to the rest of the country. Solar cells also gave China a valuable export product. The combination of economies of scale and low-cost labor made Chinese solar cells less expensive than those of its competitors and cheap enough that they could generate electricity at prices well below conventional plants, for which fuel prices were now rising steeply. Solar-cell prices, in fact, dropped steadily as designs and manufacturing techniques improved, in a manner reminiscent of the world's earlier experience with computer chips. By 2035, China was producing not just electricity but also hydrogen with solar energy; the hydrogen, piped to urban areas, was used to power and heat apartment buildings and eventually to run pollution-free buses and cars, using fuel-cell technology developed but never widely deployed in the industrial countries.

The country's huge and increasingly targeted investment in biotechnology also paid off, with development of a way to grow food in factories rather than on farms. The initial breakthrough came in 2015, and within a decade, Chinese scientists could prompt genetically engineered yeast and algae to grow many complex carbohydrates and proteins. Scaling up the process in novel, low-temperature fermenters and developing techniques to turn raw materials into a wide array of convincingly simulated rice cakes, bean sprouts, and other traditional foods took another decade.

The new biotech factories helped eliminate food shortages in the interior parts of China. Almost from the start, the fledgling industry was overwhelmed by requests to build additional plants abroad as poor countries and aid agencies struggled to meet food shortages in other areas of the world. By 2040, biotech foods were becoming a staple of many Chinese diets and a significant factor in the country's overall food supply.

These advanced technologies proved an enormous boon to China's economy, providing inexpensive energy and food, massive export earnings, and a major source of employment. As a result, China's economy continued to expand rapidly instead of slowing as expected after 2025. Not only had China increased its self-sufficiency and reduced its environmental problems, but it had also *won unprecedented international political influence, especially among energy- and food-dependent countries.*

As this scenario suggests, China may have a stronger incentive to develop and use alternative energy sources than do the established industrial countries. And even such ambitious technological steps might be within its capability: China has been steadily increasing its research efforts, has been shrewd in its technology deals with global companies, and has an undoubted ability to manage large-scale programs. Thus, given the right kind of leadership and vision, some of China's critical resource constraints and emerging environmental problems could also become opportunities. Indeed, China, lacking the entrenched private sector interests of the industrial countries, may find it politically easier to exploit emerging new energy technologies than does, for example, the United States.

Can China manage its growth by following a *Market World* strategy? Will the Communist Party eventually cede power and allow China to become a democracy, or might its insistence on political control eventually stop the process of economic liberalization? Might unresolved social and environmental problems undermine China's prosperity and threaten its stability, turning it into a *Fortress World*? Will it become a fully integrated member—indeed, a leader—of the world economy, or might it become instead a security threat to its neighbors? Can it summon the vision and entrepreneurial skill to solve its problems in ways that will benefit the world as a whole, rather than exporting its problems as pollution or excess demand? All these futures are plausible. And with a fifth of the world's population, which future China chooses will have a major impact on the twenty-first century.

Southeast Asia: The Tiger Economies

In the fall of 1997, Southeast Asia's problems briefly dominated the news: sinking currencies that triggered a worldwide drop in stock markets, economic retrenching, and a massive episode of air pollution. Yet most media coverage overlooked the region's overall track record, the best of any developing region. Economic and social progress has been rapid and widespread as country after country has adopted market-oriented economic policies and improved the health and education of its population. Compared with the situation in most parts of the world, these benefits have been shared relatively equitably; poverty has declined throughout much of the region. The original Asian tigers, exporting goods to the United States, Japan, and Europe and gradually moving up from trinkets and tennis shoes to semiconductors and cars, remain the best available prototype for how developing countries can thrive in a global market. More recently, tariff barriers are beginning to fall and intraregional trade has been growing rapidly. Even if the region's economy takes several years to recover, its long-term prospects seem very solid; the fundamental conditions for growth already exist. Indeed, except in Indonesia, where the toppling of an aging dictator and the corrupt cronie capitalism he allowed plunged the country into a period of political crisis and uncertainty, the financial crisis has brought changes to the region that will enhance its long-term prospects. These include a more democratic government in Thailand and the election of President Kim Dae Jung in South Korea, whose outspoken commitment to democracy, economic reform, and social justice already marks him as a new kind of Asian leader.

Looking ahead to the next half century, the region's most important challenges will be managing growth, adopting more democratic systems of government, and coping with its huge neighbor, China.

Critical Trends

Where does Southeast Asia appear to be headed, based on current trends? Fertility has been declining rapidly; South Korea and Singapore have reached fertility rates of two children per woman, the level needed to stabilize their populations. The region's population, now 550 million, is expected to reach 890 million (with a plausible range of 720 million to nearly 1.1 billion) by the middle of the next century.[10] Thus, demographic pressure will grow, fueling competition for resources and further stressing urban areas.

Prior to the recent economic downturn, the countries of the region together had a $2.2 trillion economy. If Southeast Asia recovers its momentum, as seems likely, output could reach nearly $25 trillion in the year 2050, under the high economic projection. Clearly, the region has enormous economic potential.

Average incomes vary widely within the region, from the near European levels of South Korea, Taiwan, Singapore, and Hong Kong to the still impoverished levels of Vietnam, Laos, and Cambodia, where market reforms are just beginning. But for purposes of comparison, imagine a typical inhabitant of the region. For such a person, in the best case, annual income could reach $38,000 by the middle of the next century—a figure higher than the comparable projection for China and second only to Latin America among developing regions. Such a Southeast Asia would be prosperous indeed. Even under more moderate midrange projections, average income would reach $18,000, and many of the region's inhabitants might attain lifestyles comparable to those in Europe today. Absent a complete collapse of the region's economies, the range given by these two figures suggests the level of development that may be plausible over the next half century.

However, Southeast Asia also faces serious environmental problems, including air pollution, rapid cutting and clearing of forests, and damage to its coastal and marine ecosystems, which are among the most diverse and productive in the world. The gravity of these problems was dramatically demonstrated in 1997 when fires set to clear forests created a 3,000-kilometer-long cloud of smoke and smog that covered the region for weeks. Pollution levels in many cities reached extremely hazardous levels, sending thousands of people to hospitals and clinics, and the near zero visibility contributed to an airplane crash and the collision of two ships. Yet environmental pressures can be expected to intensify. Energy use in the region is projected to increase fivefold in the next half century—nearly as much as in China—raising the risk of even more air pollution. Industrial activity, and hence the potential for toxic emissions, is expected to increase by nearly a factor of nine.

Some countries in Southeast Asia are rich enough to afford the cost of environmental cleanup, and many of the region's governments have shown the ability to tackle environmental problems. Wealthy Singapore prides itself on rigid enforcement of pollution laws. Indonesia, not the richest country in the region by any means, has cut back dramatically on its use of pesticides (and resultant water pollution) with an innovative farming program while still increasing farm yields. The long-term outlook is that although Southeast Asia cannot afford to ignore its pollution problems, they need not seriously constrain the region's success.

Resource scarcity could pose a more serious constraint. Much of the region's growth has been financed by sales of natural resources, especially timber and oil. But these resources are rapidly becoming depleted. Some countries, such as Vietnam, may run short of farmland for its still rapidly rising population; Indonesia may face growing conflict over access to land and forests.[11] Water is generally plentiful in the region, but shared water resources, such as the Mekong River, which rises in southern China and drains parts of Laos, Thailand, Cambodia, and Vietnam, are a potential source of conflict.

Scarcities and conflicts, if they occur, could undermine economic growth—global capital is quick to flee under such circumstances. A more serious long-term handicap may be the region's lack of advanced training for its citizens—investments in higher education have been minimal so far. Equally problematic may be the region's mostly authoritarian governments and their belief in an "Asian model" of development that finds democracy superfluous. Yet such governments have tolerated a corrosive level of corruption and, as the recent economic crisis shows, do not have a perfect track record in regulating financial systems. And will the vaunted Asian values of hard work and high savings endure as incomes rise, or will they be supplanted by Western-style materialism, as the proliferation of designer boutiques and Mercedes-Benz automobiles in the region's major cities might indicate?

These critical trends suggest that Southeast Asia's success cannot be taken for granted. But of all developing regions, Southeast Asia has perhaps the best head start toward creating an optimistic future for itself.

Chapter 12

India: A Second Independence?

IN 1997, INDIA CELEBRATED its fiftieth anniversary of independence from British rule. For the citizens of this huge country that reaches from the snow-capped Himalayas to the coastal villages of Kerala and spans arid deserts, equatorial forests, and the fertile floodplain of the mighty Ganges River, there is much to celebrate—not just the resilience of the world's largest and most diverse democracy but also its remarkable social, agricultural, and industrial progress. Life expectancy has doubled, for example, and the specter of famine no longer stalks the subcontinent; in high-tech enclaves, India's computer scientists and biotechnologists are an emerging global force.

Yet the news is not all good. Over the past half century, the average Chinese has fared far better under a ruthlessly authoritarian government than has the average Indian under democracy. Citizens of Southeast Asian countries that were no more advanced than India even a few decades ago now have average incomes ten or twenty times higher than those in India. Indeed, for some Indians, especially those living in the country's 650,000 villages, little has changed in half a century. India—and the South Asian region as a whole—has the world's worst poverty, the most widespread malnutrition, and the most extensive use of child labor.[1] Nearly half of India's adults and two-thirds of its women are illiterate. Environmental degradation is widespread and resource scarcities are growing, helping to further impoverish India's rural population. Adding to these pressures, India's population is still growing rapidly and seems likely to surpass that of China before it stabilizes, making it the world's most populous nation and one of the most crowded. Ethnic and religious conflict, and the remains of a caste system that once condemned many people to a miserable existence and still limits their social and economic advancement, mar the legacy of Mahatma Gandhi. AIDS is spreading rapidly. Tensions within South Asia—between India and Pakistan over Kashmir, between India and Bangladesh over illegal immigration, in the ongoing guerrilla war by Tamil separatists in Sri Lanka—could still erupt into deadly violence; both India and Pakistan have nuclear weapons and large armies.

But lately, India seems to be on an upswing. It is beginning to reform its state-dominated and heavily protected economy, freeing it somewhat from the country's paralyzing bureaucracy and beginning to attract foreign investment. Economic growth has averaged close to 6 percent in recent years. The country's well-educated middle class, although only perhaps a sixth of the population, is nonetheless nearly as large as that in the United States—a tempting market and a reservoir of expertise that gives India a strong potential advantage over both China and Southeast Asia. The poor and lower-caste groups are increasingly mobilized and using their new political leverage to demand more attention to the country's social needs; grassroots groups are active throughout the country. And South Asia is making progress in resolving regional tensions, even tentatively addressing the idea of a free-trade zone.

Nonetheless, far more sweeping reforms will be needed if India is to realize its potential. India's state-owned monopolies and huge central bureaucracy—the legacy of both its colonial past and its decades-long devotion to a socialist model of development—have severely constrained its development. The state monopolies that generate and dis-

India

Critical Trends

A snapshot of the future based on projections of critical trends. The graphs show the plausible future range for population, economic output, and per capita GNP—a measure of prosperity. The environmental projections suggest how much polluting activities could expand, under conditions of moderate economic growth.

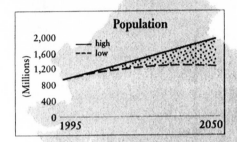

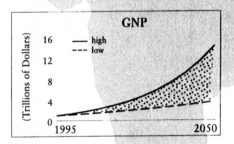

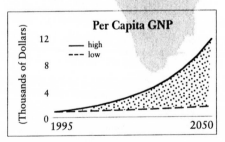

[.:.] plausible future range

Environmental Projections

Potential growth in pollution from projected increases in energy use and industrial activity by 2050, compared to 1995 levels:

Air-polluting emissions: 520 percent
Toxic emissions: 980 percent

tribute electric power, for example, have proved notoriously inefficient and environmentally destructive. So have India's natural resource agencies—the state claims all rights over India's forests, minerals, and water resources, in effect setting the government in conflict with its many citizens, who use and depend on these resources. The centralized bureaucracy still dominates decision making and stifles local initiative, yet the Official Secrets Act, taken over from the British colonial administration, shields most of the government's actions from public scrutiny.

The result has been neither the emphasis on education and relative equity characteristic of many socialist states nor the rapid economic growth and relative efficiency of most capitalist systems. Instead, India has achieved the worst of both approaches: environmental degradation and enormous inefficiency and, at the same time enormous social deprivation and disparities in access to basic services.

With a sixth of the world's population living in India, the country's success or failure will be critical to South Asia and to the world. But will economic reform go far enough to stimulate lasting economic growth and provide jobs for India's masses? Will political and social reform enable India's poor to improve their lives? Can both of these changes happen rapidly enough to prevent a disastrous downward spiral of environmental degradation, impoverishment, and population growth?

Critical Trends

India is already a densely populated country, with 930 million people in an area one-third the size of the United States. But with fertility still high, especially among the 70 percent of the population living in rural areas, the country's population is still rising rapidly and is expected to number 1.5 billion (with a plausible range of 1.2 billion to 1.9 billion) by the middle of the next century. Demographic pressures on land and other environmental resources and on swelling urban areas, already severe, will thus intensify, both in India and throughout South Asia. Indeed, given the endemic poverty of the region, stabilizing its population in the next half century may prove very difficult if current trends continue.

Despite recent signs of vigor, India's economy produces less than $4 per person per day. If economic reform were to succeed and India's economy were to grow rapidly over the next half century, it could conceivably reach the current size of the U.S. economy. If it were to grow even more rapidly—launching an economic takeoff comparable to those experienced by the Southeast Asian tigers—then India might well emerge as a global economic power, with output twice the size of Western Europe's present economy.

What would that mean for the prosperity of the average Indian? In the best case, assuming high economic growth and the lowest plausible population growth, average annual income could reach nearly $12,000—a huge increase over today. Under midrange projections, average annual income output might reach about $5,500. In the worst case, assuming spiraling population growth and economic growth so slow as to represent a virtual collapse, annual incomes might average no more than $2,000, not even a doubling of present levels. These figures suggest the range of economic development that may be plausible over the next half century.

Yet equally important may be India's social development, because here the country has lagged even further behind its eastern Asian neighbors than in economic development. Despite the country's widespread impoverishment, with 25 percent of its people living in extreme poverty, an even larger portion are deprived of even minimal social services: a third lack basic health care and half are educationally deprived; nearly 30 percent of India's children work, often as bonded laborers.[2] Indeed, by a number of measures, India and its South Asian neighbors represent the most socially underdeveloped region in the world. Sub-Saharan Africa, for example, now has fewer malnourished children and higher literacy rates than does South Asia. And although India is making some social progress, it is doing so very slowly: a U.N. agency estimates that unless conditions change, 100 years will be needed to bring India's human development up to current levels in the industrial countries. Long before then, these unremedied social problems could undercut a successful future for India.

Environmental problems could also pose a critical constraint on India's development.[3] Urban air pollution in the country's major cities is already bad, yet a projected fivefold growth in energy use could make it far worse if current patterns continue, although India is lucky in one respect: its coal, on which it depends heavily, is very low in sulfur, largely eliminating the dangers of acid precipitation. Exposure to toxic materials in factories and from industrial wastes dumped into rivers is already a serious problem, one that may increase if India's industrial activity expands nearly tenfold in the next half century as expected. Enforcement of existing pollution laws has been ineffective.

By far the most critical of India's environmental trends, however, is growing scarcity of renewable resources—water, firewood and other forest products, and fertile cropland. Such scarcities already undercut the ability of many rural communities to survive. But India's rising population and ongoing pattern of environmental degradation, if unchecked, will make the problem much worse and could undermine any chance for a successful future. India is rich in good farmland, but by 2050 it will be approaching five people for every acre of it (0.09 hectare per person), so it can ill afford to degrade what it has and must farm it more intensively. But intensive farming techniques such as those of the green revolution have not spread to many parts of the country, in part because of water scarcities and in part because the social preconditions for modern agriculture—farmers educated enough to use more sophisticated techniques, access to credit and crop advice, roads to get crops to urban markets—don't yet exist in many of India's less-developed states. Overall, India already uses more than 20 percent of its available water resources, and a new U.N. report projects that the number will exceed 40 percent—a critical level—by the year 2025. Climate change may actually increase average rainfall. If this does occur, it will be a blessing for much of India, but more intense downpours may also accelerate erosion of many of the country's deforested hillsides and Himalayan ranges.

What do these critical trends suggest for India's future? A pivotal issue is whether India will be able to lift its huge population, or most of it, out of poverty and social deprivation. If it fails to do so, it will have a hard time restraining population growth, feeding its people adequately, and preserving its soils, forests, and other biological resources. One possibility is a *Market World* future based on continuing economic reform. But would that suffice to eliminate poverty and protect the environment? If not, the possibility of human suffering on a vast scale, of a growing gulf between rich and poor within India, and perhaps increasing conflict—a *Fortress World*—cannot be ruled out. A third possibility is the development of far more fundamental social and political changes—a *Transformed World*—in addition to economic reform.

Which of these futures will India choose?

Market World: The Indian Tiger

The *"new India"* began to emerge in urban centers around the turn of the century, marked by the startling growth of young professionals with cellular phones, sleek cars, high-rise apartments, and other trappings of upscale lifestyles—India's yuppies, they were called. But they also represented the emergence of a self-confident, ambitious middle class eager to get on with building a world-class economy, an India that worked. The rising expectations of members of this new middle class and their impatience with old ways helped to solidify and extend India's economic reforms. Over the first decade of the new century, India's state gradually relinquished its stranglehold on economic activities— regulations were pruned back, state-owned companies were sold off, labor markets were freed, and domestic markets were opened to global competition. The result was an economic boom of historic proportions. By 2015, with China's growth beginning to slow, India had the fastest-growing economy in the world, and the boom continued almost unchecked for another two decades.

One factor that sustained the country's rapid growth was a construction boom as India built roads, airports, power plants, hospitals, and schools. Another was a huge drop in corruption as India's independent judiciary and investigative agencies—increasingly staffed by young, well-trained professionals—locked up thousands of crooked politicians and businessmen. Yet a third was an enormous tide of foreign investment attracted by India's booming markets, improving infrastructure, and relatively corruption-free business climate. But perhaps the largest factor was a surge in formation of new, entrepreneurial companies that sprang up to *provide goods and services. Many Indians—and many foreign observers—had feared that Indian companies would not be able to compete globally and that opening the economy would cause them to fail. Many did fail, but hundreds of thousands of new firms appeared, and, more closely attuned to the Indian market than the large global companies, they found and exploited niches, thrived, and grew. Indeed, Indian companies now rank among the most successful worldwide in such fields as software engineering, pharmaceuticals, finance, and solar energy.*

Of course, India's economic takeoff initially left behind many of the poor, and even today some poverty remains. But the construction boom brought jobs and new lives to many; others found employment in factories or the booming retail sector or as domestic help for the expanding middle class. But what helped the most, although it took a generation, was better schools—which state governments began to push in earnest as competition set in among India's states for new factories and other commercial investments—and enforcement of laws prohibiting child labor. As ten and then twelve years of school became the norm and standards improved, graduates found they could easily get jobs. Today, all but a few of India's huge population can not only meet basic needs but also enjoy consumer luxuries—one reason why the Indian economy continues to thrive.

Is such a future plausible? That depends in part on whether India can continue its economic reforms. India has yet to sell off or disband the state-owned companies, many of them inefficient monopolies, that produce half the country's output. Their sale alone would give a huge boost to India's economic growth. Still, the economy has been growing more than twice as fast as the U.S. economy, and the "new India" yuppies are indeed evident in Mumbai (Bombay), Bangalore, and other urban centers. Unlike an older generation committed to socialism and fearful of joining the global economy, this rising young generation has embraced the market. A recent World Bank study concurs, describing India as an emerging giant that is likely to become an economic powerhouse over the next twenty-five years.[4]

Moreover, India has a huge pool of well-educated, technically trained people that is proportionately much larger than those in China and Southeast Asia; India's number of scientists and engineers, for example, is second only to that of the United States. If their skills can be tapped by economic reform, these workers represent an unparalleled resource, a way for India to leap over the shortage of managers, engineers, and other professionals that hampers virtually all other developing countries. India is also staking a claim in a number of high-tech fields. It is already one of the largest developers of computer software outside the United States; it has a rapidly developing biotechnology industry; and it recently initiated a huge solar energy project—large enough to make India a world leader in solar technology.

Political support for continued reform seems likely. Despite loud voices advocating economic nationalism and seeking to keep foreign companies out, India's recent governments—five different coalitions since 1996—have not wavered. Indeed, the effect of coalition politics at the national level has been to shift more real decision making to India's state governments, a number of which are actively seeking foreign investment and have sent trade missions to tour the United States.

Still, widening economic reforms may yet falter because of the political costs entailed. Cutting chronic budget deficits will require cutting subsidies, which are popular with politicians and with those who receive them, just as are subsidies (or "entitlements") in the United States. For example, privatizing state monopolies such as electric power companies will also require abandoning the practice of providing virtually free electricity to rural farmers, some of whom may no longer be able to pump water and irrigate their fields. So the initial effect of economic reform may be to deepen poverty, and thus it may provoke strong political opposition. Nonetheless, several Indian states are attempting to break up and privatize their power companies.

A *Market World* strategy for India has one serious drawback: the strategy cannot, by itself, protect India's environment or address its overwhelming social problems. Will enough of India's population benefit from the new India? Will India's poor, now beginning to flex the political muscle inherent in their numbers, tolerate the shrinking of India's socialist remnants? Can India's social problems wait a generation or more for the benefits of a market economy to trickle down to its least advantaged? In sum, might such a strategy, whatever its initial economic advantages, lead to a more pessimistic future?

Fortress World

Rising economic growth benefited India's middle class. But the "new India" did not include most farmers, and it entirely skipped the large portion of India's population living outside the cash economy, dependent for most of their food and their livelihood on what they could grow or gather. Economic reforms, in fact, had the perverse effect of intensifying pressure on India's environmental resources—pollution increased; more rivers were dammed for power; demand for wood and paper soared. These demands increasingly uprooted rural people or destroyed their livelihoods, ballooning the ranks of the poor. With no incentive for couples to have fewer children, fertility among the poor stayed high and India's population continued to grow, putting even greater demands on the country's forests, fisheries, and soils. The burdens of preventing famine and meeting other minimum needs—politically necessary to any state government that hoped to stay in power—bankrupted several of the poorer states, forcing expensive rescue operations that increasingly drained the national treasury. Interest rates rose, and economic growth slowed.

As the number of desperate families struggling to make ends meet increased, the political struggle over India's future intensified and ethnic and religious divisions deepened. Religious riots between fundamentalist Hindus and Muslims broke out anew across the country. As villages were torn asunder by conflicts over land and other resources—often pitting landowning castes against poorer ones—violence ensued. Criminal groups, already heavily armed, grew larger and more brazen. For a while, the new India continued to soar. The new urban overclass, like the maharajahs of old, flaunted their wealth, but with television displaying their lavish lifestyles to those less fortunate, India's sense of common purpose unraveled. Populist politicians, voicing the economic frustrations of the poor, focused their anger on the well-to-do and on foreign companies attracted to India's growing market. As a result, urban mobs attacked a number of luxury apartment buildings and global companies' factories and distribution centers.

Weighed down by these problems, India never accomplished its takeoff. Foreign investment slowed, and many of India's newly rich sent their money abroad. The political consensus behind reform faltered, and the economy stagnated. Eventually, the needs of the poor began to overwhelm the country's ability to meet those needs, and when the monsoon failed, famine reappeared. By 2050, the scale of India's human tragedy was immense, with little hope on the horizon.

Could such a dismal future occur? Many of India's sophisticated urban dwellers would deny it. But the country's fisheries are declining, forcing local fishermen to abandon their traditional livelihoods.[5] India's forests are under enormous pressure, not just to provide timber but also as a source of fodder for livestock. Overpumping of aquifers for irrigation, already widespread, will eventually exhaust the supply. Many of India's soils are depleted. Indeed, India is already experiencing what Madhav Gadgil, a distinguished Indian scientist, calls an "ecological crisis," wrought by the destruction of resources needed to support its rural communities. Yet the number of people who depend on these resources will increase by 300 million in the next two decades alone.

Religious tensions are also high. India has experienced more than 8,000 religious riots in the past fifty years, and fanaticism is on the rise. In one of the most notorious incidents of religious conflict, Hindu fundamentalists burned down the Babri Mosque in Ayodhya in 1992. There is a broader culture of violence, too—struggles between castes, as well as dowry murders and other violence against women. Criminal groups operate with seeming impunity; some of their leaders even hold elected office. In part, this rising conflict may reflect the growing resource scarcity throughout rural India.

If social and environmental problems undercut economic growth, might such divisions also undermine India's democracy? In a country with seventeen official languages, marked ethnic differences, and the remnants of caste-based discrimination, India's democracy—the unifying idea of India as a nation—is the country's crowning achievement of the past fifty years. But television is bringing a sense of modern India to many once isolated villages, and the images do not necessarily strengthen common ties. The *New York Times* recently carried an interview with a twenty-eight-year-old Muslim, Mohammed Rafiq, who makes his living selling vegetables in a poor rural village in northern India. "When we sit in our homes at night, we see pictures on our television sets of the rich man's life, with cars, refrigerators, air-conditioners, everything we don't have," he was quoted as saying.[6] Attitudes among the poor, says the *Times*, may be hardening.

To avoid a *Fortress World*, India will need sweeping political and social reforms, perhaps even a renewal of basic values. One penetrating vision of such a future comes from Gadgil, who in a recent book, *Ecology and Equity*, describes "an India that might be"—a future that requires, he says, a synthesis of the Gandhian way, the Marxist utopia, and the capitalist dream.[7] I have used many of Gadgil's ideas in the scenario that follows.

India Transformed

As the economic reforms gained momentum, so did a shift in political power from the central government to the states and, gradually, even to villages. Elected village-level councils, established toward the end of the twentieth century, demanded more control over local forest and water resources and a share of the revenue raised from their sale. Some states moved rapidly toward local control, but progress was uneven until the constitution was changed, guaranteeing revenue sharing and a joint role in resource management for the local councils.

India's rural communities responded avidly to their roles as stewards of natural resources. Free from the control of state and national forest agencies, they hired their own guards, established rules regulating firewood collection, and began charging their own residents for irrigation water. As revenues flowed into village councils, some invested in new wells and other fixtures, but most spent the funds on improving local schools or establishing health clinics, buying books, or paying a teacher's salary—all services supposedly but rarely provided by state governments. In a critical case, India's Supreme Court upheld a village council's refusal to let a state forest department cut and sell timber under the village's jurisdiction.

Within a decade, decentralization and local control had brought about significant change. With both legal and illegal cutting reduced, forest quality improved. At the same time, the shortage of timber stimulated a boom in private tree farms. Village incomes were up, and local council money helped fund small-scale factories and weaving cooperatives. Social conditions in rural areas improved rapidly.

The political effects of India's rural revival were striking. Villages throughout one northern state banded together to oppose a series of large hydroelectric dams that would have flooded their lands. In statewide election, the candidate of the governing party was decisively defeated in favor of a member of the opposition party, who hastened to halt development of the dams. In a southern state, villages incensed over the poor quality of their schools elected a representative of a women's party promising to increase state spending on education and to let local communities choose their school teachers. The movement to decentralize and reform government, already well under way, became unstoppable as India's political parties competed to cut the size and power of central bureaucracies. Subsidies to businesses, to cities, and even to farmers began to disappear. New laws required extensive public disclosure at all levels of government; the laws also permitted citizen's initiatives to be placed on the ballot.

The economic effects of deregulation and government downsizing were dramatic. With village as well as urban economies growing rapidly, creating an expanding internal market, India's economic growth accelerated. India became a favored target for international investment as company after company built factories and set up distribution centers. But because of the intense local democratic activity, companies found they had to come to terms with village and district councils. Building roads or upgrading local telephone service in exchange for land and water rights became common.

Equally dramatic was the revival of the Gandhian ideal of modest lifestyles, of consuming what one needed and not just what one could. As a result, the country's savings rate soared, providing an additional source of investment even while limiting the growth of some consumer products. The country remained predominantly vegetarian in diet and thus utilized its agricultural resources efficiently.

India's economic expansion had a unique character, dominated more by basic goods than by high-profile consumer goods; overall, the country's focus was on information, services, and science-based companies—as if the country had gone directly to a postindustrial economy. Growth averaged nearly 10 percent per year for nearly two decades, and real poverty became increasingly scarce. India had become both a showcase for grassroots democracy and the ultimate Asian tiger.

Is such a future plausible? Certainly, some of its elements can be found in India today. Democracy is clearly one of India's strengths—poor people vote in great numbers and cherish their right to do so. Their increasing political activity—as reflected in the rise of caste-based political parties—is bringing enormous change. In one of the country's most populous states, Uttar Pradesh, for example, two parties representing the higher castes and the very lowest castes, known as "dalits" (formerly "untouchables"), formed an improbable governing coalition in which each would control the state for six months.[8] During the six months when the leader of the Bahujan Samaj Party—Mayawati (she has only one name), a forty-one-year-old teacher born into a dalit group—served as chief minister, she radically altered state priorities and focused state development funds on roads, wells, and village electrification that would benefit the poorest segments of the state's population. The program was so popular that the Bahujan Samaj Party's coalition partner, the upper-caste Bharatiya Janata Party, promised to continue it.

Local control of resources also has some precedents. India's forests have long been under the control of state and national governments. In the state of West Bengal, however, a partnership between the state forest department and a number of villages—now more than a thousand—has thrived. Villagers got jobs in the forest, a cut of the revenue from selling timber, and permission to gather controlled amounts of firewood, fodder, and other forest materials for their personal use in return for protecting the forest. The result? Improved forest quality, more jobs, and other local benefits; one study showed that the more than seventy different plant species gathered from the local forest were worth much more than the timber.[9]

Some Indian states—especially the southern state of Kerala—have made education and health care a priority. Today, literacy and access to health care in Kerala are essentially universal among both men and women, even though the state remains among the poorest in India. Moreover, Kerala's birthrate has dropped to about two children per family. Clearly, India has the potential for a social transformation, especially given the shifting political winds now blowing and the emerging dynamism of the country's unleashed private sector.

Nonetheless, India's economic and social transformations are mostly still to come, dependent on choices yet to be made. Overwhelmingly, India still resembles an embryonic *Fortress World*—with "islands of prosperity, oceans of poverty," in Gadgil's telling phrase. Piecemeal reform may not be enough to change this pattern or even to create a true *Market World*. But there is an emerging consensus that socialism has not worked and that economic reforms must continue. Whether such reforms and the emerging political foment can sweep aside India's problems and transform the country, as the scenario suggests, remains unclear.

Chapter 13

Sub-Saharan Africa: Transformation or Tragedy?

FROM THE SHIFTING SANDS of the Sahara to the Cape of Good Hope, sub-Saharan Africa is immense, two and a half times the size of the United States and home to nearly 600 million people. The region— what used to be called "black" Africa, although it is in fact multiracial—encompasses some forty-seven countries and a far larger number of cultures and languages. Yet notwithstanding its size and vast mineral wealth, sub-Saharan Africa is largely undeveloped: the entire region has fewer telephones than does Manhattan, the central borough of New York City.[1]

Despite its reputation as being densely crowded—indeed, overwhelmed by its teeming numbers—sub-Saharan Africa is today less densely populated than the United States. But the average woman in the region bears five or six children, so the population is growing rapidly, expanding by 2.8 percent per year. At the same time, the region's economy is smaller than that of England and over the past fifteen years has

Sub-Saharan Africa

Critical Trends

A snapshot of the future based on projections of critical trends. The graphs show the plausible future range for population, economic output, and per capita GNP—a measure of prosperity. The environmental projections suggest how much polluting activities could expand, under conditions of moderate economic growth, and how many people could face potential scarcity of water and cropland, under conditions of medium population growth.

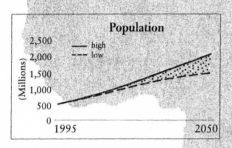

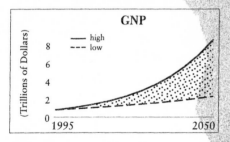

Environmental Projections

Potential growth in pollution from projected increases in energy use and industrial activity by 2050, compared to 1995 levels:

Air-polluting emissions: 530 percent
Toxic emissions: 850 percent

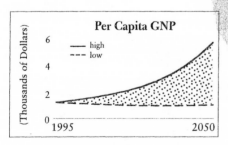

Number of people (in millions) facing potential scarcity of renewable resources, assuming no further degradation

	1995	2050
Water scarcity	43	1,100
Land scarcity	2.5	812

grown by only about 1.5 percent annually, so average income has steadily declined.[2] Poverty is widespread: the World Bank estimates that one out of two Africans subsists on less than $1 per day. Many of sub-Saharan Africa's institutions are fragile and often ineffective, perhaps not surprising in a region only little more than a generation past independence from its colonial masters. But these weak institutions have made economic growth difficult, despite the fact that sub-Saharan Africa receives more aid per capita than any other region. Transportation costs are high, and corruption is endemic. In addition, deforestation, overgrazing, and improper farming threaten the soils and other natural resources critical to livelihoods and to future prosperity in the region's primarily agrarian economies. The Food and Agriculture Organization of the United Nations estimates that as much as 80 percent of the land in Africa is threatened by degradation. In urban areas, rapid growth, deteriorating social services, and widespread unemployment have increasingly spawned crime, public disorder, and social unrest. Governments have disintegrated in a handful of countries, leading to civil war or widespread collapse of civil order and, often, large numbers of refugees.[3]

In addition to its internal problems, sub-Saharan Africa faces several external challenges. Establishment of sustained economic growth is hampered by a long-term trend of falling prices for many of the region's main exports, largely metal ores and agricultural products. A tradition of relatively high salaries in the formal economy, left over from the colonial period, makes sub-Saharan Africa relatively uncompetitive with many Asian countries, where wages are lower. Moreover, huge foreign debts resulting from reckless borrowing by past governments burden many of its economies further. With the exception of South Africa and a few other countries, foreign investment has been virtually nonexistent in recent years. Much of sub-Saharan Africa, not surprisingly, is effectively isolated from the global economy.

At the same time, however, the region's assets are substantial, and many of them are underutilized. Even after its population doubles, sub-Saharan Africa will have more arable land, water, and forests per capita than either China or India. In fact, experts think that sub-Saharan Africa could feed several times its present population if its lands and water resources were properly managed.

The region has recorded some real social progress over the past fifty years in literacy, access to education, and health care, although some of these advances have slowed or even reversed in the past decade. Although the status of women in sub-Saharan Africa is relatively low, women play a more significant role in the region's workforce than in most other developing regions. For example, in Nigeria, 36 percent of the women work outside the home; in Kenya, 46 percent; and in Uganda, 48 percent—a strong social and economic asset for the future. Moreover, sub-Saharan Africa has a vibrant informal economic sector and many small enterprises, which official economic statistics do not fully capture. The past couple of decades have seen the rise of grassroots movements focusing on health, environment, and women's status. African extended families and the habit of sharing provide a resilience that helps the region to weather adverse conditions. The African people are mobile, and large-scale migrations to areas of opportunity, to seek jobs or better grazing for livestock, are common.

Three of sub-Saharan Africa's largest countries illustrate the contradictions—the failed potential, and the bright promise—of this region. Sudan, which has some of the richest farmland in Africa and access to waters of the Nile River, has been embroiled in a decades-long civil war between its Arab north and its long-exploited black south. The war has left much of the country in chaos and poverty, cut off from foreign aid and even most medical help. The former Zaire (now the Democratic Republic of Congo) has the world's second-largest tropical forest and enormous water and mineral resources, but after three decades of disastrous, kleptocratic rule by Mobutu Sésé Séko, the country is a shambles, its roads virtually impassable, its people destitute. The third country, South Africa, has a more positive aspect. After spending decades as an international pariah for its apartheid policy of racial segregation, South Africa has emerged as both the region's—perhaps the world's—premier model of multiracial democracy and sub-Saharan Africa's foremost economic hope.

Are the xenophobia of Sudan, the mismanagement and greed of the former Zaire and present-day Nigeria, and the ethnic conflict and human misery of Rwanda emblematic of Africa's past or of its future? The answer matters. Sub-Saharan Africa is expected to have three times its present population in 2050, and if development fails, the result will be at the least a humanitarian nightmare and probably an environmental, health, and security nightmare for the world at large. Yet the future of sub-Saharan Africa is very much in doubt as the region teeters between collapse and stability. Which way the balance tips will depend primarily on the region itself, on its peoples and their leaders; but what the rest of the world does and how it interacts with sub-Saharan Africa may also play an important role.

The challenge facing sub-Saharan Africa is not just one of establishing economic reform or improving the quality of life for its people. Transcending all other issues, including population growth, environmental degradation, poverty, and economic stagnation, is governance—in the broad sense, including not just the national government but the judiciary, state and local governments, and other institutions that form African society's collective decision-making and managerial process. Better governance is the key to the region's ability to solve any of its problems. Without enforceable contracts and other laws and regulatory guidelines, markets will flounder; without functioning governments and other social institutions, no society can articulate its goals and harness its human resources to meet them. Yet sub-Saharan Africa has been plagued, more than any other developing region, by disastrously ineffective governments.

Critical Trends

Sub-Saharan Africa's population, now larger than Europe's, is growing more rapidly than that of any other part of the world. With more than half of the region's population living in poverty and with subsistence agriculture the dominant livelihood, children are a valued economic resource, both socially and economically. Access to contraception remains limited. Not surprisingly, with neither the incentive nor the means to limit it, fertility remains higher in sub-Saharan Africa than in any other region. The prospect, potentially disastrous in its implications, is that the region's population will triple by the middle of the coming century, reaching 1.8 billion (the plausible range is 1.5 billion—still more than a doubling—to 2.1 billion), even allowing for the potential effect of the AIDS epidemic. But the region cannot feed itself now. Sub-Saharan Africa imports 20 percent of its grain supply, many countries in the region rely on food aid, and nearly one-third of the population suffers from chronic malnutrition.

But if rapidly rising populations are a disturbing portent, the lack of sustained economic growth in much of the region is equally troubling. Without rapid economic expansion, prosperity will remain out of reach and many sub-Saharan Africans will sink deeper into poverty, a fate experienced by nearly a third of the region's countries over the past two decades.

Nonetheless, let us assume that rapid, sustained growth can be achieved, as indeed a handful of countries have already demonstrated. By 2050, with very rapid growth, the region's economic output could conceivably reach a level roughly comparable to that of the United States today. Even in the best case, however—with high economic growth and lower population growth—annual average income would reach only $5,700 by the middle of the next century. In the worst case, with slow or episodic economic expansion and high population growth, the figure could be less than today's $1,300 per person. Sub-Saharan Africa, the projections suggest, must run hard just to stand still and must sprint uphill to achieve even modest prosperity for its people in the coming half century. Yet the real worst case—if parts of the region experience long periods of economic stagnancy or decline—could be growing dependence on food imports and food aid, hundreds of millions of lives stunted or prematurely ended by malnutrition and desperate poverty, and the potential for widespread social chaos and violence.

The region also faces severe environmental constraints. Africa's urban areas are not yet badly polluted, despite the squalor of the slum settlements that surround them, but conditions are likely to worsen as the region industrializes and consumes more energy. Aside from South Africa, few countries in the region are likely to have the wealth or the technical skills to control pollution for decades to come. More serious is the growing scarcity of such resources as soils, forests, pasturelands, and water in underground aquifers. As noted earlier, soil degradation is widespread. Many forests are being depleted. And as populations grow, more people seek to graze their cattle on the same pastures; subsistence farm plots get smaller with each generation.

By 2050, Africa will still have a relative abundance of cultivated land—about 0.13 hectare per person, or about three people per acre, which is more land per person than China has. In theory, with modern farming methods, this is plenty of acreage to allow the country to feed itself. But in practice, the use of fertilizer, irrigation, and other modern farming methods is still rare, and thus chronic malnutrition is expected to increase by 50 percent, affecting 300 million people, by the year 2010. Moreover, except for the Congo basin and some coastal areas in West Africa, many countries will lack enough water to irrigate farmland. Some seventeen countries in the region are expected to face potential water scarcity by the year 2050. Yet to bring water from where it is plentiful to where it is scarce would require near continental-scale water projects, necessitating a degree of stability and multicountry cooperation that is lacking now.

Drought cycles are a permanent feature in several parts of the region, and climate change is likely to make such cycles more intense. Indeed, Sub-Saharan Africa stands to lose more from climate change than does any other region. A recent international study concludes that such changes as lower rainfall and more rapidly drying soils, although manageable by a modernizing Africa with effective governments, could "threaten the lives and livelihoods" of poor subsistence populations already under economic, environmental, and social stress.[4]

What kind of futures might these trends suggest? Is there a plausible path to an optimistic future for sub-Saharan Africa? That was the question pondered by African scholars and community leaders at a workshop held in 1994 in Harare, Zimbabwe, under the auspices of the 2050 Project.[5] The workshop produced scenarios and a rich store of thinking about Africa's future. I have adapted and drawn from these scenarios, so the interpretations are mine, but they build substantially on the earlier work.

Market World: Southern-Led Growth

South Africa's political transition was not lost on its neighbors. Within a decade, most of them had copied elements of South Africa's constitution and its social and economic development strategy. And as South Africa's economy began to accelerate, so did those of the whole of southern Africa, nearly a dozen countries in all. The growing stability of the subregion attracted more foreign investment, and as new highways and railroads linked countries tighter together, a common market emerged.

As southern Africa began to boom, other subregions turned to it for help. The languishing giant of the Congo bid to join the southern African common market, offering access to its mineral wealth in return for investments in roads and other badly needed transportation and communications links. Soon Angolan, Zambian, and South African companies were active all across central Africa. Uganda, already an emerging economic success on its own, led several East African countries into the common market, which, now spanning nearly twenty-five countries, was attracting serious attention from multinational companies around the world.

Although trade ties were their only formal links, these countries increasingly collaborated in other ways—putting informal pressure on neighboring countries that lagged in social reform, for example, or lending military advisors and diplomatic help to create the new nation of South Sudan, freeing it from domination by the Islamic government in the north. In more than one instance, troops from several countries intervened to overturn a coup or oust a repressive strongman. In international affairs, the southern-led group became an effective coalition, aligning members' positions and votes and thus garnering far greater clout. Although there was talk of a United States of Africa, the political coalition remained an informal one, united by a common approach.

West Africa—especially Nigeria—remained aloof from the new movement for some years. But as the southern-led coalition became undeniably successful in raising living standards, even Nigeria asked to join the common market, accepting as a condition of membership the downsizing of its army and adoption of some social reforms.

By 2050, nearly all of sub-Saharan Africa was making progress. The African common market, now nearly forty countries strong, had become a vibrant trade zone, a growing market that no global corporation could overlook. As for South Africa itself, it had become an economic giant on a global scale, the acknowledged leader of the African group of nations and a force in international affairs.

Could such an optimistic future come to pass? It may be too early to tell. But clearly, the outcome of South Africa's peaceful transition from apartheid to a multiracial democracy has been profound.

Indeed, not only has South Africa's negotiated settlement provided a badly needed beacon of hope, but also its carefully constructed constitution and new institutions, including its Constitutional Court and its Truth and Reconciliation Commission, are providing political and ethical models for sub-Saharan Africa and for the world. Moreover, the country is attempting to shift from a socialist to an open market economy while also undertaking radical social reform—integrating schools, expanding health care, and attempting to eliminate poverty. These are ambitious goals given the country's huge social needs, the imbalance in wealth and ownership of property among whites and blacks, continuing massive unemployment among blacks, and rising crime rates. How the country will fare in the future without the leadership of President Nelson Mandela, Africa's extraordinary statesman, is uncertain. Nonetheless, South Africa has strong institutions in government, higher education, and other sectors and a vigorous private sector with capable financial and industrial firms. Moreover, it benefits from a high degree of internal goodwill—so far, its people have been willing to wait for progress.

As South Africa goes, so goes southern Africa? Not only are the twelve countries that constitute the Southern African Development Community at peace, but also their economies are beginning to accelerate. And throughout the region, economic reform is favored. Privatization of mines and other government-owned enterprises is beginning; so are ambitious efforts to interconnect the region via new roads and electric power grids.[6] There is still much to do—communications are poor and red tape at borders prevents the rise of a true common market—but it is possible to imagine a regional takeoff that could bring genuine prosperity to southern Africa within a decade or two. And with a strong engine in the south—a source of inspiration, example, capital, and expertise—might not other parts of Africa gradually follow in the train?

Unfortunately, the train might just as easily run in reverse. If other parts of Africa succumb to stagnation, ever deeper poverty, political paralysis, violence and social chaos, their misery might well overflow, jeopardizing countries in the region that might otherwise succeed.

Fortress World: Collapse and Beyond

In retrospect, the early signs of Africa's disaster were already evident in the ethnic warfare and sporadic government collapses that began in the 1990s. The unsettled conditions continued into the new century and discouraged private investment. Providers of international aid, dismayed by a continued pattern of corruption and self-serving governments, increasingly put their money elsewhere. A handful of countries escaped the trend and made real progress, but elsewhere in the region economic activity stagnated and poverty deepened. With rapidly rising populations, falling incomes, decaying roads, and corrupt and self-serving governments, many Africans came close to the brink of destruction. Food became increasingly scarce—more than half the population in some countries was malnourished—and crime became the hallmark of urban areas.

The collapse into near anarchy began in West Africa as state after state foundered and then spread across central Africa and into East Africa. At one point fighting was reported in a dozen countries simultaneously. The flood tide of refugees overwhelmed the region's few stable nations and threatened southern Africa as well. The sheer numbers of migrants, along with widespread chaos and violence, overwhelmed aid agencies and made it impossible to bring in food and medicines. After a few halfhearted attempts, the western nations abandoned efforts to intervene. At the height of the troubles, more than five million people per year died of violence, hunger, and disease.

In the absence of fundamental change, sub-Saharan Africa will face growing demographic and environmental pressures and the possibility of widespread instability. From instability could come economic stagnation and, eventually, a downward spiral into widespread collapse, with the region reduced to a few heavily guarded enclaves of prosperity and stability amid a sea of chaos. What would follow? Many scenarios are possible. Would sub-Saharan Africa recover by itself with its separate nations intact, or might its nations disappear for a while, with Africa becoming a continent of villages? Would the international community intervene ' on an unprecedented scale, perhaps creating an African Development Authority to prevent further starvation, restore order, and restart development? Or might smaller-scale national or private sector interventions occur, aimed at recovering specific mineral resources, thus marking the emergence of a new colonialism? Or would the world simply turn its back on such a deeply troubled region? None of these futures could be called positive.

Yet sub-Saharan Africa's prospects are far from hopeless. Already, there are tentative signs that Africans themselves will shape a positive future for their region.

Transformed World: The Big Lift[7]

The pattern of episodic crises and occasional collapse continued into the new century, affecting first one country and then another. Almost by default, the region, apart from South Africa and a few other countries, partially decoupled from the global economy: past debts were canceled, but new loans and private investment were scarce, and export markets shrank steadily.

But the effect of decoupling was an awakening—a realization at many levels of African society that the region had only itself to rely on—and was accompanied by growing self-scrutiny and consensus for change. A new generation of pragmatic leaders appeared, determined to bring their countries into the modern world. Corruption diminished. Economic policies designed to attract foreign investment and stimulate local entrepreneurship— opening markets to free competition, reducing red tape, privatizing government-owned businesses—began to succeed here and there. The South African example proved a powerful stimulus; its businesses became a growing source of investment capital and managerial expertise for southern and central Africa. Cross-border trade expanded and flourished throughout the region. Many governments made education and other basic social services their highest priority, and some countries adopted radical decentralization programs, stimulating grass-roots initiatives and rural development.

By 2020, these reforms were almost universal. As the region's productive capacity improved, exports again found markets abroad. More global companies, attracted by wages now far below those of Asia and Latin America, built factories and service facilities in the region to serve both export and growing regional markets. Sub-Saharan Africa rejoined the global economy with a vengeance: a period of rapid growth ensued, an African takeoff that lasted several decades.

Satellite-based cellular telecommunications networks spread rapidly, interconnecting the region and providing remote rural areas with modern services, information, and entertainment. With this exposure, economic and political sophistication rose steadily, increasing the demand for new products and for improved services and better government. As the region stabilized, it devoted substantial resources to reclaiming and upgrading its agricultural base and developing indigenous energy sources. Several large-scale water projects brought many new areas under irrigation, and yields rose rapidly. By 2050, the level of agricultural production was nearly four times that of half a century earlier, and the region was exporting large quantities of food to India and the Middle East. At the same time, the geothermal energy resources of the Rift Valley had been tapped to make East Africa a low-cost industrial and manufacturing center.

In 2050, lifestyles in sub-Saharan Africa were still modest, but obtaining basic necessities was no longer a struggle for most people. There was again a sense of quality in African life and a pride in the region's growing accomplishments.

Is such an optimistic scenario at all realistic for sub-Saharan Africa? The answer can be heard in the voices of a new generation of African leaders who speak of the future in firmly positive tones. The scenario's assumptions can be challenged—that sweeping reforms and a new social consensus can be achieved, that critical environmental resources can be protected, that stability will come. But even in a region known more for its Rwandas and its Zaires, there are optimistic signs. Pragmatic, economically sophisticated political leaders are emerging, leaders who appear genuinely interested in improving their country rather than enriching themselves. There is evidence of a more confident, self-reliant attitude in a few countries that are intent less on seeking foreign aid and more on solving problems and cooperating with their neighbors. "Uganda needs just two things," says that country's president, Yoweri Museveni, one of the most outspoken of the new leaders. "We need infrastructure and we need foreign investment. . . . The rest we shall do ourselves."[8] Successful democratic elections in Ghana, Benin, and Mali and a tentative era of peace in Ethiopia and Mozambique suggest growing stability. Privatization is beginning, slowly—government agencies in charge of agricultural marketing and supply of fertilizer and seeds have been replaced by private businesses in Tanzania, Zambia, and other countries. Economic growth is improving in parts of the region, at least for now: in 1996, more than two-thirds of sub-Saharan countries grew at a rate of 5 percent.

Moreover, the region does have undeveloped agricultural potential. Only a quarter of its arable land is cultivated, with less than 3 percent of that under irrigation, and its energy potential—in oil, natural gas, and hydropower as well as geothermal energy—is scarcely tapped at all.[9] Political and social reforms include efforts to remake the judiciary systems in Mali, Zambia, and Mozambique and to democratize local governments and devolve greater authority and resources to the local level in Uganda and Ghana. Education and health budgets have reached 40 percent of government expenditures in Ghana while military budgets have fallen there and elsewhere, especially among the region's new democracies.[10] The powerful example of South Africa, the region's economic powerhouse, has given democratic reformers throughout the region the upper hand, and widespread support for South Africa's government both internally and internationally is attracting investment capital and raising interest in the whole region. Community groups are playing an increasing role, as exemplified by Kenya's Green Belt Movement, which is undertaking land and forest reclamation projects, although not all of the region's governments yet welcome such initiatives.

Still, the region's problems are overwhelming. To cite one vivid example from a piece of on-the-ground reportage: "Sierra Leone's post-dictatorship problems are almost absurd in their breadth. It once exported rice; now it can't feed itself. The life span of the average citizen is 39, the shortest in Africa. Unemployment stands at 87 percent and tuberculosis is spreading out of control. Corruption, brazen and ubiquitous, is a cancer on the economy."[11] Moreover, since this report appeared, a military coup and civil war ended Sierra Leone's attempt at democracy. In any event, the deep-seated nature of Africa's crisis means that there are no quick fixes.

Chapter 14

North Africa and the Middle East: Autocracy Forever?

OIL-RICH, ARID, AND ISLAMIC. That about sums up most outsiders' impressions of this vast and varied region that stretches from the Atlantic coast of Morocco to the Persian Gulf, from the mountains of Turkey and Iran to the desert sands of the Arabian Peninsula. Its diversity spans two major sects of Islam as well as Christianity and Judaism, many distinct ethnic groups, and profound cultural and political differences from country to country. Yet other than the occasional story about terrorism or a new provocation by Iraq, most Western media attention to this region focuses on the conflict between Israel and its neighbors. That conflict is important because it serves as a barrier to modernization and economic integration, but from the perspective of the coming half century, it is not the most important story to be told about the region, nor is it the issue on which the region's future will depend.

Does the key story revolve around petroleum and the region's vast

North Africa and the Middle East

Critical Trends

A snapshot of the future based on projections of critical trends. The graphs show the plausible future range for population, economic output, and per capita GNP—a measure of prosperity. The environmental projections suggest how much polluting activities could expand, under conditions of moderate economic growth, and how many people could face potential scarcity of water and cropland, under conditions of medium population growth.

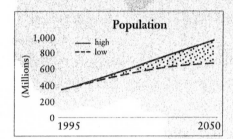

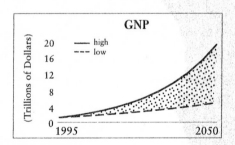

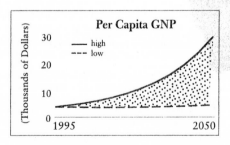

plausible future range

Environmental Projections

Potential growth in pollution from projected increases in energy use and industrial activity by 2050, compared to 1995 levels:

Air-polluting emissions: 490 percent
Toxic emissions: 630 percent

Number of people (in millions) facing potential scarcity of renewable resources, assuming no further degradation

	1995	2050
Water scarcity	155	597
Land scarcity	68	285

oil reserves? Those reserves are unquestionably vital, especially for oil-importing countries such as the United States, Japan, and most of Europe, which will increasingly depend on them. By 2010, the oil cartel dominated by Middle Eastern oil producers is expected to supply half of the world's oil—the same share it had at the time of the 1973 Arab oil embargo.[1] By 2025, just five countries in the region may produce more than 60 percent of the world's oil. Yet ironically, for the region itself, oil will play a steadily diminishing role. Long before 2025, thousands of pumping stations shading the desert landscape will fall silent as wells dry up and oil production declines in all North African countries and all but a few Middle Eastern states.

Perhaps, then, water, not oil, is the key story to be told. Water is, after all, the region's most critical resource, the one that future wars may be fought over. Within a generation, demand for water is expected to exceed by a factor of four the region's available supply. But water by itself will not determine the region's destiny either.

What will most critically shape the region's future is its lack of democratic governance or, indeed, progressive governance of any kind. Democracy's worldwide advance has gained no ground here: only Israel and Turkey have stable democracies, and except in Lebanon, no Arab head of state has gained power by democratic means for a generation.[2] More than any other region, North Africa and the Middle East have been burdened with backward-looking and autocratic—sometimes xenophobic—governments that seem incapable of successfully modernizing their countries.

One important reason for widespread autocracy in the region is the dominant role of Islam. With no other political framework, governments lean on Islam for legitimacy,[3] which reinforces inward-looking and antidemocratic tendencies. The rise of radical Islamic factions has intensified antidemocratic pressures, driving governments to intensify repression and, with a few exceptions, to turn aside political and economic reforms that might lessen their control or open their societies to greater external influences. Islamic law also underlies the region's social conservatism, obstructing needed improvements in the role of women and other social reforms.

But other factors hold back the region's progress as well. Low levels of literacy, for example, create a barrier both to political participation and to economic development. And surging populations, fueled by a growth rate second only to that of sub-Saharan Africa, guarantee increasing shortages of water and a rising demand for jobs and social services—challenges that few governments in the region seem equipped to meet. Is the region condemned to halfhearted modernization and sluggish economic growth, to continued repression, conflict, and instability? That would be a very unpromising future.

Yet recently, tentative signs of change have appeared. Egypt and Morocco are cutting budget deficits, privatizing state-owned industries, and wooing foreign investors. Some of the oil-rich Gulf sheikdoms are diversifying their economies, explicitly planning for the day when their oil runs out. In Iran, the region's most prominent radical Islamic republic, voters defied the mullahs and overwhelmingly elected a moderate-leaning president, Mohammed Khatami. Slowly, citizen's organizations that promote democracy by registering and educating voters are beginning to appear. Might the beginning of fundamental political and social transformation be under way?

Critical Trends

Throughout much of North Africa and the Middle East, female literacy is low and women face severe restrictions on their activities outside the home. Indeed, the status of women is comparable to that in far more impoverished regions, such as India. Partly as a result, fertility is high, averaging more than four children per woman. The region's population, now 350 million, is expected to more than double by midcentury, reaching nearly 770 million (the plausible range is 650 million to 935 million).

The region's economic output stands at roughly $1.7 trillion,[4] larger than that of India, despite a far smaller population. As a result, the region is hardly impoverished, although annual incomes vary widely from one country to another—in Kuwait, incomes are higher than those in Europe. But with the value of the region's oil wealth declining—real prices are now about half what they were prior to the first Arab oil embargo—and market-oriented reforms sparse, economic growth has been sluggish. If the trend continues, the region's economic output might grow only slightly faster than its population. But if the region were to grow rapidly and consistently, its output in the year 2050 might conceivably reach more than twice the size of the present U.S. economy.

What would such contrasting possibilities mean for the inhabitants of the region? In the best case, with high economic growth and low population growth, average incomes might reach $30,000 per person by the middle of the next century, higher than present U.S. levels. With stagnant growth and surging populations, on the other hand, average income might be as low as $5,500 per person, just slightly higher than at present. Much will depend on economic reforms that raise incomes and on social reforms that reduce birthrates; both may depend on polit- ical reforms not yet in prospect.

Yet even with reform, the region faces serious resource constraints and growing environmental problems. More than 60 million people live in cities with dangerously high levels of air pollution. Over the next half century, energy use in the region is expected to grow fivefold and indus- trial activity sixfold, and pollution emissions might increase compara- bly if cleaner technologies and more effective controls are not put in place.

Water pollution is also a growing problem, exacerbating an almost regionwide water crisis. Ten countries are consuming more than 100 percent of their renewable supplies, draining underground aquifers that will run dry within a generation. Even so, 45 million people in the region do not have access to safe drinking water, cities are expanding rapidly and overrunning their supplies, and there are no reserves to meet future needs. Virtually every country in the region, except Turkey and perhaps Iraq, faces potential water scarcity by the year 2050, just from the demands of growing populations.

Moreover, water is viewed with intense, nationalistic and sometimes religious fervor—the Koran calls it the source of life—making it a ready source of conflict. Egypt uses virtually all the Nile's water, but it is only a matter of time before Sudan, Ethiopia, and other upstream countries demand some of it as well. Syria and Iraq depend on the Tigris and Euphrates Rivers, yet Turkey plans an ambitious series of upstream dams and irrigation projects that will cut the flow of these rivers almost in half. Israel, which takes most of the Jordan River's water and pumps from aquifers underneath the West Bank, is now bound by treaty to share these resources with Jordan and the region of Palestine—where supplies are very low—leaving even less to meet its growing population.

Even under current conditions, by 2025 the demand for water in the region will be four times what nature can provide, and climate change is likely to make things worse, reducing rainfall and intensifying drought. Inevitably, the region must change how it uses water: it may have to import most of its grain rather than grow it or jointly manage the available water and find innovative ways to reuse wastewater. But such changes will displace farmers and transform traditional ways of rural life; they will require a shift to more modern, industrialized economies and sophisticated sewage treatment and water reclamation systems; and they will mean expanded trade in and dependence on the global economy—changes that threaten current political alignments. Thus, environmental—as well as social and economic—pressures point to the need for political change. The alternative may be instability and conflict, with countries going to war over water or convulsing from internal pressures.

What, then, might these trends suggest for North Africa and the Middle East? Is a *Market World* future conceivable? For most of the region, such a scenario would stretch plausibility, as things now stand; I don't attempt to develop it. Too many governments fear the loss of control that might come from opening their countries to the West and joining the global economy, and fundamentalist Islamic factions would strenuously—perhaps violently—oppose such policies. Not even autocratic rulers can lightly dismiss such opposition; assassination of political and religious leaders figures prominently in the region's history. Yet without such a global link, countries in the region will find it very difficult to modernize their economies, gain access to advanced technologies, and create enough jobs for their rapidly expanding and youthful population.

Is revolutionary political change a more likely alternative? In large part because autocratic governments have not tolerated political opponents, radical Islamic factions have become the only apparent vehicle for change. Indeed, the longer governments tolerate corruption and fail to provide social services and the conditions for job-creating growth, the stronger is radical Islam's appeal and the greater is the likelihood of open conflict. The result might be a continuing wave of instability and violence, a *Fortress World*—perhaps coupled with outside intervention by Western powers to seize some of the major oil fields.

Still another, more hopeful, possibility would be evolutionary but still fundamental political and social change—a *Transformed World*—that could set free the region's latent economic dynamism and build on its strong universities and lengthy scientific tradition. The following scenarios are my adaptations of material originally prepared for the 2050 Project.[5]

Fortress World: Everlasting Firestorm

The second Islamic revolution began in Algeria, where the refusal of the military-backed government to accept a victory by the Islamic parties at the ballot resulted in a violent campaign of terror. The government refused to compromise with the militants, and the situation disintegrated into open civil war. Before the war ended with the fall of the government in a bloody slaughter of "infidels," nearly a hundred thousand had died and many others had fled the country.

As poverty and unemployment worsened and water grew scarce, the frenzy of violence swept into Tunisia, then Libya, and finally Egypt. There, after two years of mass demonstrations, mounting terrorism, and economic stagnation, the president resigned in order to prevent further bloodshed and invited the Muslim Brotherhood to form a government.

Violent overthrows of governments were not limited to North Africa. Members of the conservative House of Saud, bickering among themselves over succession, were openly condemned by several elderly Islamic clerics, and the regime started to come apart. Within months, to the horror of the United States, a radical opposition appeared—supported, as it turned out, by senior army officers—and took power with only modest bloodshed.

Most of the Saudi princes were killed or allowed to flee abroad. Even where revolution did not succeed, governments tightened repression and limited outside contact. In Turkey, the military again temporarily seized control to thwart rising fundamentalist sentiments, but the situation remained tense.

To protect its access to oil, the United States quickly formed alliances with Kuwait and the United Arab Emirates—where a standard of living approaching that of Europe tempered the xenophobia and conservatism of the rest of the region—and established military bases, gradually increasing its military presence on the peninsula. Although the new Islamic governments proved willing to sell oil to the West, thus keeping their commercial relations with the rest of the world largely intact, internal changes in those countries were dramatic. Strict adherence to Islamic traditions was required: women were compelled to stay at home, and crime was brutally punished.

One deleterious outcome of the changes was an exodus of professionals and their families, stranding the more sophisticated parts of the region's economy. Oil production gradually dropped, as did commercial trade with the rest of the world. Over time, other difficulties appeared. Unemployment deepened. Despite drastic penalties, black markets flourished. Population growth surged, and acute shortages of water and food became commonplace. Disagreements between Sudan and Egypt over the Nile escalated to a brief war. Egypt won decisively, but not before Sudan destroyed the Aswān Dam—the ensuing flood killed nearly a million people—and with it Egypt's ability to feed itself. Even under near famine conditions, however, most of the Islamic regimes proved unwilling or unable to ask the West for help or even to tolerate international aid organizations. Protests and violence, never entirely absent, became a constant feature of the region, as did abrupt shifts of government control from one faction to another, each one usually harsher and more conservative than the last.

Other than a concern about oil supplies and, for Europe, a constant flow of illegal immigrants, the world came to ignore the region. Even oil became less of an issue as vehicles powered by fuel cells—running on hydrogen from natural gas or from solar refineries—replaced the internal combustion fleet; by 2020, these vehicles constituted more than half of the automobiles on the road worldwide. As the region of North Africa and the Middle East turned in upon itself, conditions became increasingly miserable.

Could such a scenario happen? At present, Syria, Iraq, and Libya forcibly exclude fundamentalist Islamic movements; Egypt and Tunisia repress them; and Saudi Arabia, Morocco, and Sudan have tried to pre-empt them. Only in Jordan, Yemen, Lebanon, and Kuwait is there some limited accommodation, and only in Iran and Afghanistan have such movements gained control of governments. These movements are more than just a religious phenomenon. With no other avenue for dissent in these societies, radical Islam has attracted some of the brightest and most capable young people; in most countries, Islamic movements have moderate as well as militant or xenophobic factions. But as the movements gain in strength and yet are denied any peaceful outlet, they appear to be turning more to violence. Such conflict is already under way in Algeria, where violence has reached horrendous levels, and dissent appears to be growing in Tunisia, Morocco, and Egypt and in Saudi Arabia, where the government's invitation of foreign troops onto Saudi soil during the Gulf War brought charges that the regime dishonored the country. Even in secular Turkey, there is rising interest in religious matters coupled with growing disgust at government corruption.

Most governments in the region seem to be on a no-win trajectory. On the one hand they refuse to engage Islamic groups in democratic processes, reinforcing radical tendencies; on the other, they are doing little to improve economic and social conditions, potentially draining away support. Apart from limited attempts in North Africa, there is little sign of vigorous reform: state socialism mitigated by black markets still rules, not market economies. And, as noted earlier, the Islamic culture itself makes rapid progress extremely difficult in raising the status of women or effecting other social changes that touch the family. Nor does there appear to be much official cognizance of the water crisis. Both Saudi Arabia and Libya, for example, are draining irreplaceable underground reservoirs to grow wheat—in Libya's case after pumping the water across a desert at great expense—when the wheat could be far more cheaply bought abroad. Many countries subsidize water prices, encouraging wasteful use. In short, the region is rapidly drifting in a dangerous direction.

If radical Islamic regimes do come to power, as in Iran and Afghanistan, they are likely to be captive to their religious rhetoric and hence poorly prepared to improve the status of women, slow population growth, or modernize their economies.[6]

Is there a more optimistic alternative for the region?

Transformed World: Islamic Breakthrough

Although few would have predicted it at the turn of the century, there is no denying that North Africa and the Middle East are now dynamic players in the global economy. Social and economic historians still dispute the causes of the transformation. Some point to the region's increasingly well-educated youth, who rapidly adopted modern economic aspirations and approaches, creating an entrepreneurial boom even while retaining moderate Islamic beliefs and social traditions. Others focus on political events—the success of the peace process involving Israel and its neighbors and the rapid integration into the Israeli economy of the new state of Palestine and, to a lesser extent, of Jordan. With the welfare and income of Palestinians skyrocketing, even Syria abandoned its hostility and, with Lebanon, joined what became the Palestine common market—now one of the world's centers of high-tech startups and innovative technology.

But most observers point to the charismatic leadership and farsighted statesmanship of Safir Ibn Abdul Aziz Al Saud, a Western-trained economist and a grandson of Saudi Arabia's King Fahd. Safir's undoubted religious piety gained him the backing of key religious leaders and, in 2007, the Saudi crown. King Safir ascended to the throne just as the international oil market tightened. Unlike his predecessors, he refused to increase production: oil prices doubled within a few years and continued to rise for much *of the next two decades. At the same time, Safir, while keeping control of vastly increased oil revenues, reduced his government's spending dramatically by privatizing and liberalizing the economy, even reducing military expenditures. Then, carefully citing precedents in Islamic law and the social obligations to take care of neighbors spelled out in the Koran, he began the Islamic Development Program, fueled with what was eventually more than $100 billion of Saudi money per year. The money built schools and health clinics, trained teachers and doctors, introduced modern agricultural techniques, financed loan programs for small entrepreneurs, and created venture capital funds to stimulate industrial enterprises. The program began first in Saudi Arabia and in neighboring Yemen and Jordan but soon spread to Iraq, to Egypt, and across North Africa. Its success gave Safir extraordinary influence throughout the region, enabling him to mediate a number of disputes that might otherwise have turned to war. Safir eventually cajoled Kuwait and some of the oil-rich sheikdoms into joining the effort, which is widely credited with helping to eliminate poverty and create the social conditions for the region's economic takeoff.*

Admittedly, there is relatively little evidence of enlightened and moderate leadership within the region. Violence regularly stalls the Middle East peace process. But the region does have numerous strengths. The Arab world has a long tradition of shrewd trading; markets are not a new idea there. The region has some excellent universities and boasts a scientific tradition longer than Europe's. And if the global trends of supply and demand for oil cited earlier hold true, prices will rise sharply and the value of the region's oil might easily double or triple within a few decades. The extra cash, if invested, could become a powerful font of venture capital for the region. Likewise, if the technical expertise and entrepreneurial experience of Israel could be tapped throughout the region—to use scarce water resources more efficiently, to create new export businesses—then rapid growth is not out of the question. Some new ventures have already begun to take advantage of the region's proximity to the huge European market, growing fresh fruit and vegetables for export and developing resorts to attract tourist dollars.

For all the exhortations against the "Great Satan," much of the region's anti-Western rhetoric seems designed primarily for internal consumption, to distract people from the failures of their authoritarian regimes or as part of the internal debate between fundamentalists and proponents of modernization. Most of the region's people, it seems clear, are interested primarily in improving their living standards. Western products are avidly sought after, even such seemingly unlikely objects as Barbie dolls—reportedly a favorite among children of middle-class families in Iran.

Even female literacy, although still low in many countries, is improving. In Iran, women are already playing a significant part in the economy and, more slowly, in political matters, though they must carefully put the traditional chador over their business suits when they leave home. Saudi Arabia, faced with labor shortages, is loosening restrictions on women in the workforce as a way of easing reliance on foreign workers. Women already make up more than a third of the medical students in several North African countries. Thus, changes in the status of women may not be as impossible as they seem on the surface.

Some political changes are also evident. A number of countries are beginning to hold elections at the local level, and some have slightly eased controls on the press and on citizen's groups. Still, clear signs of democratization and genuine political reform are scarce; secret police, censorship, and other tools of autocratic regimes continue to hold sway. International pressures for reform are notably absent—most Western governments tend to back the status quo rather than support political liberalization.[7] So fundamental change will be slow, if it comes at all. But given the region's volatile political makeup and growing social needs, time, like the sand in an hourglass, may be running out.

Chapter 15

Russia and Eastern Europe: Transition to What?

THE VASTNESS THAT IS RUSSIA comprises the largest land mass of any nation on Earth. From the borders of Europe to the frozen steppes of Siberia to the Pacific coastline of Vladivostok, Russia reaches nearly halfway around the world. The country has extraordinary natural resources—40 percent of the world's reserves of natural gas; 30 percent of its standing timber; a quarter of its coal, diamonds, and gold. Although shorn of its Soviet-era empires and with just half the population of North America, Russia commands attention: in any map of the future, it looms large.

Yet Russia and the countries of eastern Europe are undergoing wrenching transitions, shifting from centrally planned to market economies and from the Soviet Union's authoritarian grip to democracy. For the Russian people, the shock has been particularly severe—not only have they lost the cradle-to-grave social welfare and predictability of life that marked the Communist system, but also their nation's super-

Russia and Eastern Europe

Critical Trends

A snapshot of the future based on projections of critical trends. The graphs show the plausible future range for population, economic output, and per capita GNP—a measure of prosperity. The environmental projections suggest how much polluting activities could expand, under conditions of moderate economic growth.

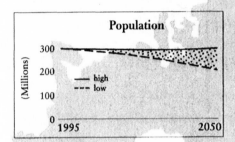

Population

(Millions)

300
200
100
0

—— high
- - - low

1995   2050

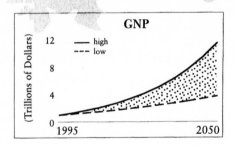

GNP

(Trillions of Dollars)

12   —— high
      - - - low
8
4
0
1995   2050

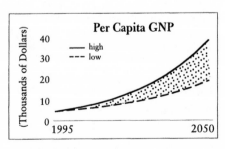

Per Capita GNP

(Thousands of Dollars)

40   —— high
30   - - - low
20
10
0
1995   2050

plausible future range

Environmental Projections

Potential growth in pollution from projected increases in energy use and industrial activity by 2050, compared to 1995 levels:

Air-polluting emissions: 180 percent
Toxic emissions: 240 percent

power status has evaporated, replaced by precipitous economic decline, rampant crime and corruption, and soaring economic disparities. Yet both Russia and eastern Europe have advantages that set them apart from most developing regions, including nearly universal literacy, an industrial infrastructure, and (especially in Russia) a large and capable scientific community; most of the countries of eastern Europe have strong historical and cultural ties to western Europe.

The economic and political transitions have been undeniably rocky and are in no way complete. Russia, for example, lacks adequate housing. Transportation, critical in such a huge country, is still rudimentary—Russia has only one-fiftieth as many roads per capita as western Europe. And although market reforms have begun in nearly every country in the region, capitalism, in the words of energy analysts Daniel Yergin and Thane Gustafson, is still "entangled in the ruins of the command economy."[1]

Moreover, the region suffers from perhaps the worst environmental degradation in the world, the result of decades of uncontrolled pollution, careless industrial practices typified by the Chernobyl nuclear reactor disaster, and misuse of natural resources, such as the massive diversion of water for irrigation that has devastated the Aral Sea. Health problems have reached extreme levels—one in five Russian children suffers from chronic illness, and life expectancy has plunged, sinking to an average of just fifty-seven years for Russian men.[2] As the *Economist* recently put it, Russia has "Africa's subsistence economy, Pakistan's corruption, Brazil's wayward congress, Italy's mafia . . . and a Communist Party all its own."[3] Because the iron fist of Communism was preceded by the autocracy of the tsars, the Russian state has so thoroughly dominated both the economy and the lives of its citizens for so long that creating an independent judiciary, an independent industrial sector, a stable democracy, and a strong civil society will not be easy. For example, the abrupt end of totalitarian control has allowed organized crime to flourish. The status of women, protected to some degree under Soviet rule, has plummeted. And relations between the Russian Federation and the independent republics that were formerly part of the Soviet Union remain unstable. The 25 million ethnic Russians living in these republics—many of them in Kazakhstan, the largest of them—provide both an excuse for continuing Russian intervention and a potential source of large-scale migration to Russia.

Whether the transition succeeds—and the trajectory it follows—is important not only to the region but also to the rest of the world. A prosperous, stable, democratic Russia could play a powerful role in world political affairs; with its immense store of natural resources, Russia would be a natural trading partner for an economically integrated Europe on the one hand and for Japan and developing Asia on the other. In contrast, a hostile and authoritarian Russia—especially one that was economically weak or that sought to reestablish its empire—could be a source of continuing instability and security threats.

Critical Trends

For its size, Russia is lightly populated, with about 150 million people. The seven largest countries of eastern Europe—Belarus, Bulgaria, the Czech Republic, Hungary, Poland, Romania, and Ukraine together have a comparable population. Russian women are not having many children; birthrates are down. In addition, average life spans are declining in both eastern Europe and, especially, Russia—a result of collapsing health systems, hazardous pollution, and, perhaps, the stress of the traumatic transitions now under way. As a result, populations are expected to decline sharply by the middle of the next century, to 131 million in eastern Europe (with a plausible range of 110 million to 156 million) and to just 114 million—nearly a 25 percent drop—in Russia (the plausible range is 97 million to 142 million).

Economically, both transitional regions are still moribund. Economic activity and currencies have plunged, especially in Russia—at the end of 1996, its economy was half the size it had been in 1989. Average annual incomes are about $4,500 per person, higher than in Southeast Asia but lower than in Latin America.[4] In a few countries, growth has started again. If the transition to market economies succeeds, economic growth could conceivably be quite rapid, given the region's well-educated workforce, relatively low salaries, proximity to the European market, and, in the case of Russia, enormous natural resources. Indeed, the World Bank recently identified Russia as one of five major emerging economies likely to play a significant role in the global market of the future.[5] If that forecast comes true, Russia's output might conceivably reach almost the current size of western Europe's economy by 2050, and eastern Europe's economy might grow to a similar size. If more moderate growth prevails, consistent with an incomplete transition and perhaps periods of instability, output for these regions might climb only slightly higher than current Japanese levels.

For the average Russian or eastern European, rapid growth would bring genuine prosperity—average annual incomes could reach $37,000-$40,000, twice those in western Europe today. With more moderate growth, average incomes might still reach $29,600 in Russia and $26,000 in eastern Europe.

Yet prosperity may well be tempered by polluted environments; degradation is already severe and is not likely to improve dramatically in the near future. If economic reform continues, the least efficient industrial plants are likely to close down or be replaced, eliminating some of the worst sources of pollution, as has already occurred in the former East Germany. Still, as economic activity increases, energy use is expected to nearly double and industrial activity could almost quadruple; without careful controls, this increased activity might generate additional air pollution and toxic emissions. How these conflicting trends—closing existing plants but building new ones and increasing the number of cars and trucks—will work out can't be determined, but given the region's health crisis, pollution trends may have significant social and political repercussions.

With declining populations, neither eastern Europe nor the Russian Federation faces potential scarcity of land or water. Indeed, both have huge agricultural potential—Ukraine's rich soils, farmed as effectively as those of Iowa, could feed much of Europe. Moreover, prospective changes in Earth's climate could, on balance, benefit the region—potentially decreasing precipitation in parts of eastern Europe but lengthening the growing season in northern Russia.

These trends suggest a wide variety of possible futures for eastern Europe and Russia. Eastern Europe's destiny, in particular, depends not only on its own efforts but also on what happens in its powerful neighboring regions—whether the economic integration of western Europe succeeds and expands eastward to incorporate the region, whether Russia becomes a stable and open market economy or a hostile and unstable neighbor. So the critical question within the region is the future of Russia. The scenarios for Russia that follow are adapted from a detailed study and a family of scenarios prepared by Daniel Yergin and his colleagues.[6] I omit a *Transformed World* scenario because even a *Market World* depends on a successful political and economic transition that is, in effect, a profound transformation of Russian society.

A Russian *Market World*

Although President Boris Yeltsin nudged Russia toward democracy, it was not until 2010 that the democratic process seemed secure. It was, to be sure, democracy with a distinctly Russian flavor, with a powerful presidency and a state that played a large role in the economy and in the lives of ordinary Russians. But stable political parties began to appear, helped in part by the political activity of an emerging business class and of a new generation of Russians who had been children when the Soviet Union collapsed. Through these parties, the Russian people demanded more effective government.

The economy expanded during the period of political consolidation. But another decade passed before Russia's transition to a rough version of a market economy was complete. Eventually, though, Russians gained secure property rights, a stable currency, and more reliable methods of taxation. The government succeeded in suppressing the more violent types of criminal activity and enacted legislation to reform business and financial activities. Corruption remained, but judicial authorities began to uphold the laws vigorously and fairly. Small business activity surged.

After Russia resolved its dispute with Japan over the Kurile Islands, Japanese investment in the Russian Far East and Siberia expanded rapidly. European and U.S. corporations, attracted by Russia's vast natural resources and large domestic market, poured money into oil, natural gas, and other ventures. Even more important, Russians stopped hoarding their capital in foreign banks and under the mattress and began investing it at home. By 2020, the Russian economic miracle was well under way.

Exports surged as Russia's revitalized manufacturing plants and skilled but relatively low-wage workers found markets around the world. New trade agreements with the expanding European Union, increasingly dependent on Russia's oil and natural gas, gave impetus to the trend. As economic growth boosted incomes, pent-up consumer demand for housing, appliances, cars, and other goods sustained the boom; the economy expanded rapidly for more than two decades. As the state's share of the economy declined, subsidies were phased out and many uncompetitive companies were downsized or closed. Some of those that survived proved to be world-class competitors. Gazprom and several Russian oil firms joined the ranks of the world's major energy companies, and the Russian language was increasingly heard in the financial districts and business hotels of the world.

Is such a future feasible? Certainly, the centralized economy of the Soviet era was so inefficient, and so focused on maintaining the Soviet Union's massive military machine, that huge improvements are possible. And there is no doubt that ordinary Russians yearn for a better life. Yergin argues that the most underutilized resource in the Soviet economy was "management, entrepreneurial talent, and innovation."[7] Already new entrepreneurial activity is surging in Russia, even if some of it is devoted to robber baron capitalism, capturing assets from the state. Nonetheless, the country still lacks the social and legal framework for a *Market World*—secure property rights, modern laws governing commerce, protection from the vicious Russian mafia. Without these, economic growth will be hobbled. So political transformation is crucial.

To many observers, President Yeltsin's reelection in 1996 signaled a consensus among the Russian people to go forward with the economic and political transformations. And economic reform, although proceeding by fits and starts, has moved surprisingly swiftly. In a breathtaking leap, Russia has privatized most of its industry; indeed, the state-owned sector now accounts for more of the national economy in Italy than it does in Russia. A few newly privatized companies show signs of becoming powerful national corporations, especially in such areas as banking, oil, and gas. The country's stock markets and other institutions of a market economy are still flawed, but they have appeared with remarkable speed. The service sector, virtually nonexistent in Soviet times, is expanding rapidly to meet consumer needs. Some Russian businesses are voluntarily adopting U.S.–style accounting procedures to make themselves more attractive to foreign investors.

The enormous diversity of the Russian Federation, with its eighty-nine different republics and other districts and regions, guarantees an uneven transition: some regions have adopted reform wholeheartedly, whereas others are still run by Soviet-style local bosses. A few "autonomous" regions such as Chechnya have tried to go their own way entirely. These differences are not just geographic: the new Russia is hardly an equitable place, as the business elite get rich while pensioners and many in the Russian military sink into penury. Some 22 percent of Russians now live below the official poverty line. If the welfare and living standards of most Russians do not improve in coming years and corruption rises, public reaction could push the country back toward a more authoritarian society.

Fortress World: Return of the Russian Bear[8]

The Russian democratic experiment did not last long, outliving President Yeltsin by only a few years. His successor proved weak and unable to forge the political consensus needed for continued economic reform. Despite continuing subsidies to many large businesses, unemployment began to rise. The government ran short of money and could not pay pensions or military salaries for months at a time. With most agricultural land still tied up in moribund cooperatives and money for fertilizer and equipment lacking, more and more Russian farmers abandoned the land; during the exceptionally harsh winter of 2005, food shortages swept through many Russian cities, necessitating an embarrassing appeal for food shipments from abroad.

The following spring, anger at the continued political stale-

mate, widespread corruption, and worsening living conditions triggered street protests in Moscow. The authorities attempted to repress the demonstrations, killing a few of the demonstrators, but their acts unleashed pent-up emotions: Russians poured into the streets, looting stores and posh restaurants. Ultranationalist groups and die-hard Communists, seeking to exploit the situation, urged the crowds to overthrow the government and restore the Soviet empire. Faced with such unrest, the president ordered the army to intervene. But the troops refused to fire on the demonstrators; indeed, some of them deserted and joined the mobs, resulting in pitched battles in the streets. Ordinary life came to a halt, and the unrest spread to other parts of the country.

With the president indecisive and the government increasingly paralyzed, the defense minister—a tough former paratroop commander seasoned in the Afghanistan campaign and known for his patriotic views and contempt for politicians—flew to Rostov and assumed direct control of the southern theater command. He quickly organized an assault by crack troops and seized control of transportation and communication links. At the same time, he appealed to other major military units for support. Most declared their allegiance. With no effective military opposition, Russia's new leader soon took control of the country, ruthlessly suppressing any civilian dissent. To the alarm of the west, the Russian Bear was back.

The government that emerged was authoritarian but quite different from that of the Soviet era. Run by a small junta of military commanders, the regime relied on the military and the secret police, in close alliance with a few captains of industry. Its ideology was vague—based not on Communism but rather on Russian nationalism and law and order. Overall, the new government was militantly anti-Western. It curbed the legislature, censored the press, purged the mafia, and arrested members of the new business class it considered traitors. Western companies abandoned most activities in Russia. But the government made no real effort to return to a command economy, and after a few years the underground economy returned, followed by larger-scale private enter-

prise. What emerged eventually was a type of market economy but in a centralized, nationalistic, and authoritarian form. Even now, in 2050, Russia remains relatively poor compared with the West— a hostile and resentful nation that poses a continuing security threat to its neighbors.

Is this version of a *Fortress World* plausible? Certainly, many of the discontents that drive it are evident in today's Russia. Moreover, prior to 1990, Russia had no experience with democracy—an authoritarian streak runs deep in the nation's character. Economic reform, although well started, could easily stall without strong leadership. If *Market World* fails, the accumulating social and economic strains could lead to protests and disorder. Whether the military would seize control is uncertain—it has tended to stay out of politics—but there are few other institutions in Russia capable of resolving a fundamental dispute between the Russian people and their government.

In addition, Russia will face enormous strains in coming decades. The country cannot afford its present welfare system. Schools, roads, and other essential infrastructure are deteriorating. Separatist tendencies in many regions of Russia and growing decentralization of authority make it more and more difficult for Moscow to govern this sprawling country. Aging and poorly designed nuclear reactors still supply a significant part of Russia's electricity, especially in heavily populated regions, and Western nuclear experts do not rule out the possibility of another disaster like the one at Chernobyl. The Russian AIDS epidemic, so far mostly ignored, is being propelled by a surge of prostitution, a transient population, and the virtual collapse of public health systems, and thus it could swell to major proportions. Chinese influence over— and possibly increasing direct interests in—parts of the Russian Far East could provoke a clash between these two nuclear-armed powers.

In short, the hopeful economic and political transitions now under way in Russia are fragile and will remain so for some time to come. Whether they continue, and what path they follow, will depend on the choices that Russia's leaders and its people make.

Chapter 16

North America, Europe, and Japan: Leadership or Stagnation?

THE INDUSTRIAL REGIONS OF THE WORLD dominate today's global economy, their output significantly larger than all developing and transitional regions of the world together. Moreover, these regions develop virtually all advanced technologies, provide a home base for most of the world's transnational corporations, and finance most international investment—so they also dominate key sources of wealth for the future. Even if economic growth slows, average incomes in these regions seem likely to at least double and could more than triple over the next half century, making the regions even more prosperous.

Beyond their sheer wealth, these regions provide their citizens with a quality of life that is the envy of the world. They boast many well-protected natural treasures—from the Alps to the Grand Canyon to Japan's Inland Sea—and even in urban areas, environmental conditions are generally improving. With stable democracies, political freedom is now taken for granted.

North America, Europe, and Japan

Critical Trends

A snapshot of the future based on projections of critical trends. The graphs show the plausible future range for population, economic output, and per capita GNP—a measure of prosperity. The environmental projections suggest how much polluting activities could expand, under conditions of moderate economic growth.

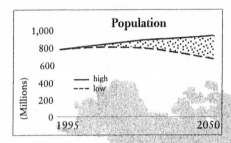

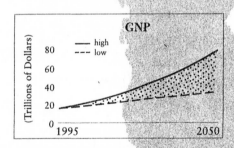

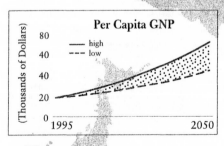

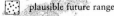
plausible future range

Environmental Projections

Potential growth in pollution from projected increases in energy use and industrial activity by 2050, compared to 1995 levels:

Air-polluting emissions: 150 percent
Toxic emissions: 180 percent

More auspicious circumstances are hard to imagine. Of all the world's regions, North America, Europe, and Japan have the fewest internal constraints on their futures—and the most to lose if the larger world turns desperate, unstable, and polluted. Surely, then, one might think, the nations of these regions will reach out to the rest of the world and use their enormous economic and social advantages and their command of capital and technology to lead the world toward more hopeful trajectories.

Yet all three regions seem beset by uncertainty about their future and often are nearly paralyzed by debate over domestic social and economic issues, unable to make needed reforms. Just a few years ago, the United States was preoccupied with widespread job insecurity and a ballooning national debt; worries persist about the social and environmental consequences of free trade. Europe continues to struggle with alarming levels of unemployment—more than 12 percent for the region as a whole, nearly 20 percent in Spain, concentrated among the young. Japan has yet to emerge from a prolonged period of economic stagnation.

Other social and political problems within these regions also demand attention: growing economic disparity and continued inner-city decay in the United States; the separatist leanings of French-speaking Quebec that threaten to tear Canada apart; the struggle over economic integration and the reemergence of hate groups and near-fascist political movements in Europe. All three regions must face the problem of paying for generous retirement and health care promises for aging populations, a problem that will become acute early in the twenty-first century.

Might these societies—preoccupied with domestic problems, resentful of growing waves of illegal immigration and the need to support peacekeeping missions in far-off places, fearful of their ability to compete economically with more vigorous, low-wage regions—turn inward and stagnate? Might they fail to lead?

Such a strategy would be self-defeating because as trade expands, even the continued prosperity of the industrial regions may depend more on the markets, and hence the stability, of developing regions. Moreover, the industrial regions cannot improve their citizens' overall welfate—as opposed to wealth—by focusing entirely on internal concerns. Over the past twenty years, for example, some thirty new infectious diseases have appeared, mostly in developing regions, posing a growing health threat to rich nations as well.[1] Moreover, an inward-looking approach would virtually guarantee a more environmentally degraded world in the future because these regions contribute disproportionately to global environmental problems. They remain the major source of carbon dioxide emissions that are changing the atmosphere, and their demand helps to deplete marine fisheries, hasten the logging of forests, and pressure other biological resources. And without leadership from the industrial regions to find solutions to such global problems, the burgeoning growth of developing regions will worsen them.

Can the industrial regions find the political will to solve these problems? Will they catalyze the evolution of a global civilization? Or is their hour on the stage of history past?

Critical Trends

The present population of North America, Europe, and Japan together is approximately 800 million—about 14 percent of the world's population—and is expected to increase only slightly, to 835 million, by 2050 (the plausible range is 685 million to 950 million). These figures hide conflicting trends: the population of North America is expected to rise by nearly 30 percent, to 384 million, over the next half century, whereas those of Europe and Japan are expected to decline by about 10 percent.[2]

These regions represent an immense concentration of economic activity, with combined output of nearly $18 trillion in 1995. By 2050, even with modest growth, output might reach $55 trillion. For the citizens of these industrial regions, average annual incomes might reach $62,000–75,000, and could go higher. By any measure, these will be very wealthy societies with average family incomes near $200,000, seemingly well able to afford additional environmental and social reforms—and also to help poorer societies, either from altruistic impulses or as investments in global stability and security.

Despite economic growth, energy use and material consumption per person are rising only slowly in these regions.[3] Nonetheless, if current trends are not checked, energy use and emissions of carbon dioxide could increase by 40 percent over the next half century.

How might these trends play out in the next half century? All three regions have highly developed market economies, so a *Market World* scenario is hardly speculative—it is largely an extrapolation of the present situation. But since all countries in these regions intervene in the market now to a greater or lesser extent, one might argue that a strict *Market World* scenario would imply significant economic reform, especially for heavily regulated Europe and protected Japanese domestic markets, requiring major social and political adjustments. Nonetheless, I won't discuss a *Market World* scenario further, since it simply reflects the conventional wisdom.

More interesting are the alternatives to *Market World*, two radically different but plausible scenarios. First, a *Transformed World* that assumes these regions remain open and innovative societies that use their wealth and other advantages to foster economic and social change domestically and to exercise continuing leadership globally. This is a hopeful vision of vigorous, confident societies acting with enlightened self-interest to create a better world. But given existing internal social and political challenges as well as threats, a *Fortress World* cannot be ruled out in any of the industrial regions: a scenario that assumes these regions turn inward, accompanied by growing rigidity and resistance to change, economic stagnation, and abandonment of world leadership. This is a vision of more selfish, fearful societies and a darker future.

Both scenarios turn on the interaction between the industrial regions and the rest of the world. For simplicity, I treat all three industrial regions together, although it is certainly possible that North America, Europe, and Japan will make different choices and take different paths toward the future.

Transformed World

The defining debates were different in each region, but cumulatively they helped to shape a new society. The initial concerns were economic. In Europe, the concern was whether or not to go ahead with economic integration—adopting a common currency, adding central European nations to the European Union. In the United States, it was whether to extend the North American trade zone to Chile and eventually to the rest of Latin America. In Japan, it was whether to open internal markets to global competition. Despite much hesitation and resistance, all three regions opted for openness. But the decisions ultimately required difficult social and political changes.

Some of the changes were driven by economic policies. In the United States, the forging of political support for trade deals required more social efforts to help those who lost jobs as a result. In contrast, as economic integration accelerated in Europe, that region gradually scaled back its generous social welfare benefits— despite strikes and protests, especially in France—and lightened the regulatory burdens placed on companies. In Japan, the opening of markets required dismantling the government–industrial linkages known as Japan, Inc., and adopting wider political reforms as well. But the result of these shifts in all three regions was more rapid economic growth.

Environmental concerns also forced change. As evidence of a change in Earth's climate accumulated, public opinion shifted in favor of taking action. Initial agreements to make modest cuts in emissions of greenhouse gases were strengthened, gradually but inexorably raising the prices of gasoline and electricity.[4] Gas-guzzling vehicles became socially unacceptable, and new cars averaged nearly eighty miles per gallon. Although coal mines were shut down, many new jobs were created in the booming fuel-cell, solar-cell, and wind energy industries. Energy use plummeted. At the same time, as governments in the industrial countries used their revenue windfalls from energy-related taxes and sales of emission permits to lower social security and other employment taxes, labor costs came down and employment surged, especially in Europe.

Lower taxes proved so popular—and the logic of reducing pollution by boosting industrial efficiency so compelling—that many countries also raised their levies on many kinds of natural resources, taxing such raw materials as metals, minerals, forest products, and even water in addition to energy. That extended the revolution in industrial efficiency: recycling rates for many materials soared. Europe's industrial "take-back" laws, which required manufacturers to reclaim and recycle most durable goods at the end of their useful life, were widely copied. At the same time, there was a pronounced decline in the consumer culture of the rich world, especially among younger people: "low-impact" lifestyles, vegetarian diets, and antimaterialist ethics gained a wide following. With less demand for raw materials and energy, pollution and waste-disposal problems in the industrial countries declined sharply.

In all three regions, the shift to more open economies and, eventually, more open societies mirrored fundamental changes taking place in the nature of economic activity. By the year 2015, commerce on the Internet exceeded $8 trillion. It was often difficult to say in which country this "frictionless capitalism" was occurring. Nonetheless, North America, Europe, and Japan benefited as the primary suppliers of the computers and telecommunications equipment on which the worldwide network depended, and they also benefited from the remarkable efficiency of the "cashless" electronic commerce that became the standard in these regions.

The more flexible economies created new companies and new jobs at unprecedented rates. The resulting new employment opportunities helped many of those who had been displaced by shifting trade patterns or pushed off welfare rolls.

Social activism also helped bring fundamental social change. In the United States, it was the urban renaissance and, in particular, social programs led by black churches and other community groups that really began to break up that country's underclass. A sense of prosperity and economic security among most Americans also prompted more generosity toward those left behind and provided political support for urban renewal efforts. Local governments, often combining welfare, anticrime, and urban renewal funds, began to channel grants through church and other community groups. Major corporations provided money and management talent and began to site new facilities to support the urban rebuilding. A new bottom-up emphasis on improving community schools and providing training for the skills demanded by the new economy began to show results. Slowly, U.S. society began to share its new wealth and expanding opportunities with all its people. Similar efforts in Europe helped to reclaim the "lost generation" of young people who had endured years, sometimes a decade or more, of unemployment. With urban revitalization, cities again became attractive, exciting places to live, greatly slowing suburban sprawl.

The effects of a more open economy were especially marked in Japan. The private sector increasingly abandoned lifetime employment guarantees and, slowly, began to open its executive ranks to women and foreigners, if only to keep up with the foreign companies operating in Japan. Younger Japanese (both men and women) were less and less willing to accept the old social contract and all it implied and are eager to be more self-reliant. Genuine political reform gradually followed, as did the emergence of a Japan that was less insular and more aware of and responsive to the concerns of its Asian neighbors—and more capable of international political leadership.

In Europe, the integrated economy began to hum. Before long, unemployment ceased to be a concern to governments in the region. Social benefits were more modest, but higher prosperity was, for most people, an acceptable alternative. The turn toward more flexible and more open economies, though painful and politically controversial at first, also made it easier to push European integration eastward, eventually all the way to the borders of Russia.

The United States continued to provide a kind of social laboratory, an incubator of innovation, for the world.[5] By the third decade of the twenty-first century, its efforts to reinvent government at all levels began to pay off. The new approach emphasized efficiency and flexibility, with widespread use of information technologies—flattening hierarchies, privatizing and decentralizing the delivery of services, and developing new problem-solving structures. As the social efforts that brought about the urban renaissance broadened their focus to address the problem of making an increasingly multicultural society really work, the country began to experience the economic advantages of diversity: continued high innovation and an ability to work with, market to, and compete in virtually every culture on the planet. In particular, U.S. companies proved to have significant advantages over their European and Japanese competitors in Latin America and Asia— regions that had contributed the most to North America's continuing influx of immigrants and were increasingly well represented in its business and professional classes.

In all the industrial regions and in the more prosperous developing countries as well, there emerged a widespread conviction that it was now within the grasp of a global civilization to eliminate poverty, much as earlier international efforts had eliminated smallpox and polio. The result was a burst of new energy and new approaches harnessing humanity's collective wisdom and compassion to solve this problem and other new challenges as well—from restoring the blighted parts of Earth to planning the eventual human colonization of Mars.

This scenario argues that open economies and a willingness to accept change initiate a virtuous cycle: more open societies, more innovation, growing tolerance, spreading affluence, and decreasing poverty. The use of markets to advance social and environmental goals, the scenario suggests, is a key element of a successful strategy. Experience supports the claim. Already, Europeans buy smaller and more fuel-efficient automobiles than do Americans—and within Europe, Italians buy more efficient cars than do Germans—because European and especially Italian governments have systematically levied higher fuel taxes, a powerful market incentive. A novel market-based approach to reduce sulfur dioxide pollution—a primary cause of acid rain—has succeeded in the United States past all expectations.

Many economists argue that Europe's lack of flexibility and tightly regulated economy are already making the region uncompetitive in the global economy and that the region will have to either alter its approach or face decline.[6] In the United States, strong environmental regulations have not compromised the growth of the economy or the ability of firms to compete globally, nor would tax policies or other redistributive mechanisms that would share the country's prosperity more fairly.[7] And a Japan with more open and competitive markets—and a more open and responsive political system—would still be able to retain the advantages of its largely equitable society and its consensus approach to important social choices.

But will U.S. society, or the societies of Europe and Japan, follow such a trajectory, or will they instead turn inward, resist change, and focus only on preserving what they have? If so, might that trigger, instead of a virtuous cycle, a vicious cycle of decline—economic stagnation, more rigidity of thought and resistance to social action, and growing intolerance and divisiveness?

Fortress World

The U.S. economic boom continued for a few years into the new century but did not lead to social progress. The failure to extend free-trade agreements to South and Central America was ascribed to a growing fear of job loss by unions representing low-skilled workers and the openly isolationist sentiments of the religious right, which had become an increasingly dominant force in the Republican Party. Innovation slowed and recessions reappeared.

Welfare reform did remove poor people from government rolls, but it left a lot of them on the street. The result was even sharper economic divisions between rich and poor as well as huge increases in the numbers of abused and abandoned children, in crime and the inner-city drug trade, and in the militancy of poverty groups and other advocacy organizations. Demand for private security services grew rapidly.

Despite booming economies in Latin America and Asia, many desperate people from those regions tried every means possible to reach the United States. The political debate over immigration intensified. One presidential candidate promised to lower quotas for legal immigration and to "fence off" the country's southern border. But it was a new cycle of urban riots, including gang-led raids on middle-class suburbs that alarmed many U.S. citizens and sparked a call for stronger law enforcement and harsher penal measures. The violence also led to a massive increase in gated communities with armed guards. Evidence also surfaced that Mexican and Colombian drug cartels had essentially taken over a major U.S. airline, a trucking company, and several other major corporations, using them to ferry and distribute drugs. With polls showing that many Americans found the world a dangerous place, the president announced a high-level commission to explore the feasibility of constructing an electronic shield around the country's perimeter to detect illegal movements of people or substances.

European political deadlock continued into the new century. Economic union had been achieved, but economic reform had not. The European Union's complex regulations, designed to harmonize members' different national laws and practices and to achieve collective social and environmental goals, discouraged business innovation. Subsidies to newer and poorer members kept taxes high. All these factors, added to the growing burden of retirees and the need to support large numbers of unemployed— 15 percent of the labor force—combined to make European economies increasingly uncompetitive. With low-priced imported goods from Korea, China, and other more efficient manufacturers swamping European markets, pressure to protect companies and jobs became overwhelming.

The ranks of the unemployed were a fertile recruiting ground for populist politicians building right-wing parties. The politicians crusaded against foreigners, who they accused of taking European jobs and importing crime, and against the European Union itself, which they blamed for the economic decline. Europe still seemed a haven for many in the poorer countries of south-central Europe, the Middle East, and North Africa, and illegal immigration continued to rise despite all efforts at interdiction. As the economic crisis deepened, ultraright parties increasingly attracted mainstream voters who were fed up with political paralysis and drawn to the parties' xenophobic and antigovernment message. By 2015, these parties had became a force to contend with in most European countries.

Two events heightened Europe's sense of anxiety, gloom, and isolation. Yielding to pressures from both right and left, the European Union adopted stiff tariffs on a wide range of imported goods. The move proved disastrous, triggering a trade war with the United States that rapidly escalated; Japan and many Southeast Asian economies gradually followed the U.S. lead. Europe, dependent on trade, entered a deep recession. Russia threatened to shut down the pipelines supplying natural gas to Europe unless it was granted an exception to the tariffs, and the European Union backed down. The United States, sensing victory, threatened to pull out of NATO and remove the last of its troops from Europe if the tariffs were not dropped for it and its partners, too. European resolve crumbled and the tariffs were rescinded, leaving Europe embarrassed and angry.

The second event was the Islamic fundamentalist uprising that swept across North Africa, triggering a mass exodus of people fleeing the violence and the strict imposition of Islamic dress and behavior codes. Flotillas of small boats overwhelmed Spanish and Italian naval patrols, even though many were turned back. Middle-class families paid thousands of dollars to be smuggled into Europe through other routes. The news media dramatized the influx and the leaky borders of the southern European Union states. Over the next year, ultraright parties were swept into office, often as part of coalitions, in a number of countries. They demanded—and most parliaments were only too ready to approve—the reimposition of internal border controls within the European Union and the establishment of other security measures, such as new identity cards. Fortress Europe increasingly abandoned its liberal ideas and retreated into itself.

With economic growth stalled in Europe and anemic in Japan and the United States, economic leadership effectively passed to Asia, where the Chinese economy continued to boom, and to Latin America, which proved increasingly self-confident and innovative. Industrial production in these regions and trade among them were now rising more rapidly than were production and trade anywhere else in the world.

Without U.S. leadership, the late-twentieth-century effort to gain international agreement on measures to protect the climate went nowhere. By 2020, however, the effects of an altered climate were inescapable. Despite international recriminations and political posturing, however, the world continued to burn ever larger quantities of oil and coal. The industrial countries showed no signs of changing their consumption patterns—indeed, some U.S. politicians went on record in support of "an American's God-given right to as much oil as he or she can use." Most people seemed to believe that climate change, like the loss of coral reefs and most of the large game animals on the plains of Africa, was inevitable, that environmental degradation was out of control and beyond society's ability to contain it.

Pessimistic scenarios are depressing. But they are cautionary tales, not predictions—they need not happen if societies take steps to alter their trajectories. Is that possible? Are there reasons for optimism about the global destiny?

PART V

GLOBAL DESTINIES

Chapter 17

Choosing Our Future

WHICH WORLD LIES AHEAD in the twenty-first century? Will the future bring a *Market World*, in which widespread prosperity, peace, and stability come from economic reform, technological innovation, and the integration of developing regions into the global economy? Or will it bring a *Fortress World*, in which the rich get richer but large portions of humanity are left behind, the environment is irreversibly degraded, and conflict, violence, and instability are widespread? Or is it possible that social and political change, driven by both enlightened leadership and grassroots social coalitions, will lead to a *Transformed World*, in which power and prosperity are more widely shared, in which markets serve social and environmental as well as economic goals, and in which basic human needs are met nearly everywhere?

I suggest that both optimistic and pessimistic futures are fully within the range of possibility, given present long-term trends. And although these trends can in principle be reversed, many of them have powerful momentum.

The optimistic trends include the spread of market-based economic reform, which may well accelerate economic growth and, over the next half century, bring prosperity to much, perhaps most, of the world's population.

Another such trend, the rapid pace of technological innovation, may make possible cleaner and more efficient industrial processes and less environmentally harmful lifestyles. The interlinking of the global market may allow developing regions to jump directly to these advanced technologies—everything from cellular phones to fuel-cell cars to village Internet links—accelerating their development.

Finally, broad social and political trends, from increased literacy and improved health care to gradual improvements in the status of women, the spread of democracy, and the decentralization of government, are also positive.

Despite these optimistic trends, however, society also faces pessimistic trends that pose serious challenges to a successful future. These include rapidly rising populations in some regions and declining populations in others. Sub-Saharan Africa's population may triple; that of North Africa and the Middle East may more than double; and the population of India may increase by more than half, making it the most populous nation on Earth. At the same time, Russia's population may shrink by 25 percent and those of Japan and Europe may increase by 10 percent or more, intensifying the economic and social problems of these aging societies.

The global pattern of economic growth has been uneven in recent decades, with rapid growth in a few countries and stagnation in many others. Even with growth in developing regions, however, gaps in income between rich and poor countries appear likely to widen, creating enormous international disparities. The switch to market economies may increase economic disparities within many countries as well.

Rapid industrialization has the potential to markedly worsen air and water pollution in developing regions, multiplying health risks and discouraging tourism and foreign investment. If present trends continue, rising global use of energy will intensify what is already a serious threat to the stability of the world's climate. At the same time, pressure on water supplies, fertile soils, forests, and fisheries—from rising populations and from growing industrial and urban demand—seems likely to further degrade many of these natural resources, perhaps irreversibly, and hence to impoverish the large populations that depend on them.

The combination of demographic, economic, and environmental trends and an increasingly borderless global economy seems likely to intensify threats to the security of individuals and the stability of communities, raising the possibility of increased migration; more crime, conflict, and violence; and perhaps chaos and collapse in some countries.

What will be the result of these conflicting trends, positive and negative? To what kind of future will they lead? I believe that the outcome depends on the choices human societies make in coming decades. We have it within our power to shape the future by our individual and collective actions, to choose which world our children and grandchildren inherit. Indeed, I am reasonably optimistic about the outcome.

How might we choose our future? I don't pretend to have all the answers needed to create a better world. But I do think we know how to fix some of the problems, to turn around many of the trends that appear to be taking the world along worrisome trajectories. There is no shortage of ideas, no lack of opportunities, if we can summon our collective will. A full discussion of even a small fraction of these opportunities would fill more than another book, and my purpose here is simply to suggest that it is possible for human societies to create more positive futures for themselves. But let me give a few examples that illustrate the hopeful innovation and promising social ferment already under way as well as some new, largely unexploited opportunities for social initiatives—examples that I hope will stimulate further thought and action.

The "Greening" of Global Corporations

U.S. and European environmental groups are finding a surprising new set of allies—major international corporations. Not long ago, I witnessed the hard-boiled executives of a major U.S. forest products company earnestly seeking the help of a group of young environmentalists dedicated to saving the world's forests. Global competition, the executives explained, meant that they needed to differentiate their products from those of similar companies in Scandinavia and Asia. They believed that with prices converging, care for the environment—in their case, forest stewardship—would in the future become a major factor in consumer choice about which brand of paper towels or toilet paper to buy. To prepare for that competition, they needed to know what the rules would be, so they wanted the environmentalists' help in determining what practices would constitute "green" forestry. The company, in effect, was asking for advice on how to improve its environmental performance and for help in setting global standards—not because its managers had suddenly become environmentalists but because they saw such changes in their practices as essential to profits and long-term success. Nor were they offering a one-way partnership. In return for the environmentalists' help, the executives were willing to help persuade other major U.S. forest companies to support the proposed new standards.

Such partnerships are multiplying, and a number of global companies, alone and in industry-wide groups, are voluntarily taking steps to lower their environmental impact or to endorse proposed environmental laws. British Petroleum, for example, has broken ranks with other energy companies and publicly acknowledged the need for companies to adopt measures to protect the climate. Moreover, corporations are coming under increasing public pressure to meet social as well as environmental expectations. Companies that sell fashionable lines of clothing or sport shoes, for example, are finding it necessary to rethink their practices and to avoid suppliers in developing countries that exploit child labor or do not provide safe working conditions.

Still, only a comparative handful of global companies have committed themselves to environmental and social goals as well as profits and have made fundamental changes in their behavior and corporate strategy; the "greening" of the private sector is a tentative trend, not an accomplished fact. But if a few leading companies succeed with such strategies, market forces will very likely compel others to follow. And might not global companies—because they are global—rapidly propagate these attitudes and practices around the world? A few are trying to do just that.

Transforming Industrial Society

Even if global companies are willing, however, they have to respond to the demands of the marketplace—to what consumers are willing to buy. Consumers, in turn, are extremely price sensitive. But pricing is not entirely in the hands of manufacturers and merchants; many prices are shaped in part by national policies that establish economic incentives. Italy, for example, places a high tax on gasoline; not surprisingly, Italians demand smaller, more efficient cars than do Americans. Both in India and in the western United States, governments subsidize irrigation water; so farmers use a lot of it, even growing crops—such as rice in California—that are water-intensive and would otherwise be foolish to grow in an arid region and thus contributing to water shortages.

Broadly speaking, most countries subsidize the use of natural resources, either directly or through their tax codes. The U.S. government, for example, subsidizes mining, logging, and cattle grazing on federal lands; production of oil, natural gas, and coal; and food production (through farm and water subsidies). Yet these activities are a major cause of pollution and other environmental problems—and, of course, low prices encourage consumers to buy more of these products based on natural resources than we otherwise might. So the policies are at cross purposes with environmental goals.

At the same time, most industrial countries tax employment, both directly through social security taxes on workers and employers and indirectly through income taxes. Yet such taxes—especially those that employers must pay—can be a powerful disincentive to hire more workers. Thus, these policies, however necessary for generating government revenue, are at cross purposes with social concerns about unemployment and job creation.

Suppose that industrial countries decided to change their tax and subsidy policies—and hence the economic incentives that shape markets—so as to support rather than work against long-term social and environmental goals. If natural resource subsidies, many of which in the United States date to the opening of the West a century ago, were gradually phased out, then over time the demand for natural resources would lessen and so would environmental pressures. If the money saved were used to reduce employment taxes, that would encourage job creation.

Such proposals are controversial, but the basic principle at issue is clear: why not employ markets to help achieve social and environmental goals by realigning market incentives with those goals? Studies suggest that such realignment offers not only environmental and social benefits but also economic benefits, increasing industrial efficiency and perhaps accelerating economic growth. Indeed, some European countries such as Sweden are going further by raising taxes on energy while reducing income and employment taxes. Over a decade or two, such an approach could gradually transform industrial societies, helping to turn around potentially dangerous environmental and social trends.

The Power of Information

Perhaps even more powerful than economic markets in influencing human behavior is the marketplace of public opinion. Most people care about what others think of them; so do must corporations and governments. Thus, policies and laws that require the disclosure of certain actions can have a major influence on behavior.

Starting in the late 1980s, a new U.S. law required companies to publicly disclose their emissions of several hundred toxic chemicals, published by the Environmental Protection Agency as the Toxic Release Inventory and updated annually. Many environmentalists found the sheer volume of potentially hazardous pollutants—all completely legal under existing environmental regulations—staggering. So did the chief executive officers of many of the chemical companies and others responsible for the emissions. Almost immediately, company after company announced voluntary cleanup efforts, and a decade later, the volume of toxic emissions has declined dramatically.

The same principle—regulation by disclosure—is behind the labels that inform U.S. consumers about the ingredients in food products or the efficiency of new cars or air conditioners. Might similar labels also be used to certify that wood products did not come from virgin forests or that garments were not made by child labor?

Disclosure might be essential to combat emerging security threats. If companies that make weapons or other dangerous technologies had to disclose, at least to the government, everyone they sold them to, that would go a long way to prevent such weapons from ending up in the hands of terrorists or rogue states such as Iraq. And suppose that a country plagued by corruption, such as Mexico, passed a law requiring banks and other financial institutions to disclose all financial transactions of public officials—might that reduce bribery? If governments themselves make public more information about their own actions, might citizen pressure demand better performance? In an era of global media coverage and widespread access to the Internet, public disclosure of information may provide an extremely powerful means of encouraging constructive behavior.

The Rise of Citizen's Groups

Citizen pressure can come from individuals, but organized community groups, environmental organizations, and other voluntary institutions of all kinds seem to be emerging as an important social and political force. For one thing, there are a lot of them. For another, such groups often seem more efficient and more nimble than governments.

It is possible that such groups will provide an effective way to deliver social services, expanding the role of traditional charities. Church-based social agencies in the United States, for example, often have better track records in dealing with poverty and drug rehabilitation than do governments, and some state and local governments are beginning to support their efforts with public funds. Internationally, a number of governments are experimenting with using citizen's groups as a conduit for development aid.

A more important role for citizen's groups may evolve from their growing influence in setting social and political agendas. Environmental groups have long had a powerful influence in the United States and in Europe, lobbying for new laws, filing court actions, and providing information to the media. Now, citizen's groups are beginning to play an international role as well. When a number of such groups recently approached the United Nations about negotiating a treaty to ban land mines, the response was that the effort would take a decade. The citizen's groups were not willing to wait that long. Organized as the International Campaign to Ban Landmines (ICBL), they worked outside U.N. channels to persuade the world to take action. This remarkable effort—which earned the ICBL and its coordinator, Jody Williams, the Nobel Peace Prize for 1997, was carried out by a hodgepodge coalition of more than 700 separate citizen's groups around the world, linked together by electronic mail. In less than two years, with the help of the government of Canada, they managed to generate enormous publicity; enlist public figures such as the late Diana, Princess of Wales, to help; coordinate successful negotiations for a formal treaty; and persuade virtually all major governments (except those of China, Russia, and the United States) to sign the treaty.

Such instances suggest that citizen's group are expanding the ways in which societies govern themselves. Could this new phenomenon help societies come to terms with needed transformations?

A New Age of Philanthropy

Citizen's groups are not the only new and hopeful phenomenon of our time. Wealthy individuals and philanthropic foundations are also emerging as a powerful social force. In 1997, with the United Nations nearly bankrupt because the United States Congress has refused to honor past commitments, Ted Turner, founder of cable television network CNN, announced that he would give $1 billion to support U.N. efforts. Some months later, financier George Soros, long a supporter of governments struggling to make the transition from Communism to market economies, announced that he would give an additional $500 million to Russia—an amount comparable to official U.S. aid for that country.

These dramatic gifts do not by themselves amount to a new activism of wealth, although Turner has promised to challenge other wealthy individuals to compete philanthropically. But the gifts do underscore an unprecedented accumulation of private wealth, much of which is likely to swell the ranks of traditional philanthropic foundations. With assets of $8 billion, the newly created David and Lucile Packard Foundation, formed by one of the founders of the Hewlett-Packard Company, a computer manufacturer, is just the first of a series of high-tech and media fortunes that are expected to create $1 trillion in new philanthropic capital in the United States alone over the next few decades.

Might philanthropy help finance a social transformation in the industrial countries and accelerate development in poorer regions? Already, such money is a primary source of support for the citizen's groups described earlier. Could philanthropists also pursue a more activist approach by encouraging (and offering to finance) government reform, helping cities try out new educational approaches, and supporting access to medical care and the Internet in poor communities and villages?

Accelerating Development

Opportunities abound to head off some of the most threatening trends in developing regions. For example, accelerating the education of women and providing the poorest segments of society access to contraception and health care—actions within the capability of nearly all developing countries—would go a considerable way toward slowing population growth. Targeting programs that provide economic opportunity to the same segments of society would do much to reduce the worst poverty.

But even more creative possibilities might arise if developing and developed regions work together. One way to implement a climate treaty, for example, would be to allow companies required to reduce emissions of greenhouse gases to do so anywhere in the world if they so choose. The atmosphere, after all, doesn't care whether carbon dioxide comes from a refinery in New Jersey or a coal-fired power plant in China—it's the cumulative effect that counts. So if a U.S. company could more cheaply reduce emissions at a site in, say, China and still get credit for it, investment in more efficient technology might be accelerated in developing countries. Indeed, estimates are that such an approach could eventually funnel hundreds of billions of dollars per year into developing countries, accelerating their development and helping to stabilize the climate at the same time.

Indeed, whether via the climate treaty or by other novel mechanisms, it seems clear that the world needs some new ways to recycle capital and share wealth between industrial and developing regions to help ameliorate growing disparities and to improve the choices and the prospects facing developing societies. Internally, most *countries* already do this to one extent or another, through graduated income taxes, welfare programs, or other kinds of transfer payments. But internationally, the only such mechanisms are foreign aid and private corporate investments in developing countries. The volume of foreign aid, whether government to government or through such institutions as the World Bank, is simply too small to make much of a difference to the destinies of most developing regions, and such top-down efforts, via governments, are often not successful, especially where governments themselves are weak. The volume of private corporate investment is rising, but it goes to a very limited number of countries and not at all to those that are poorest and most in need of help.

In effect, we have a global economy but not adequate mechanisms to cope with global problems, whether accelerating development or heading off currency crises or even combating global crime. The volume of international currency transactions is now more than $1.25 trillion per day and rising rapidly, as is the volume of international trade and flow of digital information across international borders via satellites and undersea cables—and none of these global flows of money, goods, and information are taxed at an international level. In effect, the global economy gets a free ride, while global needs go unmet. Might it not be possible to tax the global economy directly to meet such global needs, perhaps to create a global development fund or other creative new mechanisms that recycle capital? Even a very small global tax—far less than one percent—could generate a very large fund, given the scale and growth rate of the global economy. Could we not use such a fund to link together and finance community groups and hence to provide a far more bottom-up approach to assisting development, village by village and urban slum by urban slum—in effect, helping communities take care of their own development? Could not we use the skills and wide reach of global corporations to assist this process by, in effect, hiring them (directly or by establishing appropriate market incentives) to help provide far wider access in developing countries to health care services or basic agricultural advice or small-scale loans than weak governments themselves can provide?

The opportunities are many. Can such hopeful ideas be put into practice? There is a wealth of studies that suggest they could be, but politically it depends on what society decides to do. Ultimately, it depends on us.

Summing Up

And what is the human prospect? I am optimistic that human societies will find the will and the creativity to overcome the constraints and challenges they face and to establish a trajectory into the future that bears more resemblance to a *Transformed World* than to a *Fortress World*. But I have no illusion that creating such a future will be either easy or inevitable, nor do I expect that all regions will follow the same trajectory. In fact, if I live long enough to see much of the next fifty years unfold, I expect mostly to be surprised at the course of events: human society and this complex system of a world that we live in are capable of infinitely more possibilities than I can conceive of.

That we can't predict the future is not terribly important. What is much more important is that we can *shape* the future: that we try to envision the future we want and then set about making it happen. But some strategic thinking is essential. Humanity is no longer just another passenger on planet Earth. The sheer numbers of people and the scale of the human enterprise are now such as to have a lasting, perhaps irreversible, environmental impact; our capacity for destruction and the potential size of the human disasters now possible are equally large. The constraints of a finite planet and of human poverty and ignorance are real. Just as parents struggle to teach their children to think ahead, to choose a future and not just drift through life, it is high time that human society as a whole learns to do the same.

What Can We Learn from the Scenarios?

Market World illustrates the power of markets as an engine of economic growth and increased opportunity. But the scenario also illustrates the barriers that exist to fully unleashing that power. In some of the poorest countries, for example, the rule of law, effective governance, and other institutional preconditions for markets don't fully exist; until they do, prosperity or even a reduction of poverty is likely to prove elusive. In many developing and transitional countries, centralized decision making and government-owned companies still dominate the economy; without economic reform, markets cannot function effectively. Barriers exist even in the industrial regions: Japan's protectionist policies and Europe's rigidly regulated labor markets distort their economies, for example, causing high prices in Japan and high unemployment in Europe.

At the same time, the comparison of *Market World* and *Fortress World* makes clear that markets cannot do everything. Under strict *Market World* conditions, environmental degradation would increase, as would economic disparities between rich and poor; illegal migration and other social problems also might well fester and get worse. Moreover, a global economy increasingly needs some kind of global regulation—or an unprecedented degree of cooperation among national authorities—to deal with global crime, prevent financial instability, set technical standards, and resolve trade disputes. And how should the world deal with states that fail, as did Somalia, or that become a security threat to others, as has Iraq, or that simply wither away, as did Mobutu's Zaire? *Market World* provides no answers.

Transformed World does offer tentative answers to these quandaries. It suggests what may seem like radical changes in policy and improbable forms of social action. For example, redistributing land from the wealthy to the poor by breaking up large estates in regions such as Latin America sounds radical indeed. However, such land redistribution has been successfully carried out in South Korea, in Taiwan, and—in a different form—in China, where it proved to be the cornerstone of rapid rural development and helped jump-start economic growth. Or take the idea that citizen's groups can transform social conditions far more effectively than can governments, an idea that sounds idealistic until backed up by the tangible successes of efforts led by black churches in U.S. inner cities, by community development groups in Indian villages, and by women's groups in many societies.

Transformed World also suggests the need for continuing political reform, perhaps combined with decentralization, that would bring about more genuine democracy and allow more people to participate in the decisions that directly affect their lives. The scenario proposes a rethinking of the embedded financial incentives that condition both individual and corporate decision making in ways that are needlessly harmful both to the environment and to social goals—subsidies that encourage wasteful consumption of natural resources, for example, or taxes that discourage the creation of new jobs.

Which World? On-line
A HyperForum
on the Future

A HyperForum is an on-line discussion tool that uses the visual, interactive, and linking capabilities of the World Wide Web. Readers are invited to visit the *Which World?* HyperForum site on the Internet at http://www.hf.caltech.edu/WhichWorld. Visitors to the site will find synopses of the global and regional scenarios from this book, charts showing critical trends, and interactive tools designed to let them explore alternative futures. Links will connect interested parties to additional books, reports, and data pertinent to a discussion of the future.

The purpose of this HyperForum site is to enable students, educators, scholars, and interested members of the general public to explore issues related to society's future—issues such as the assumptions underlying different world views; the significance of long-term trends; the interaction of economic, social, and environmental factors in shaping the future; and implications of critical public policy issues. The site will host one or more extended on-line discussions about *Which World?* During those periods, the discussion or comment tool will be turned on and visitors to the site will be invited to participate in the discussions. In the course of on-line discussion, participants may also encounter others with similar interests and will have the opportunity to make further connections, so as to enlarge and enhance the community of those interested in our common future. To make full use of the site's capabilities, participants will need Java-enabled browser software such as Netscape Navigator 3.0 or Microsoft Internet Explorer 3.0.

HyperForum was developed jointly by the California Institute of Technology, the Rand Corporation, and the World Resources Institute, with funding by the John and Mary R. Markle Foundation. The *Which World?* on-line HyperForum site is made possible by support from the John D. and Catherine T. MacArthur Foundation.

Appendix

Regions and Projections:
The Details

The regions used in this book as units of analysis, although they correspond approximately but not precisely to those used by demographers and geographers, are those that seem to make sense from an economic and cultural perspective. They comprise three industrial regions, two transitional regions, and six developing regions. A few countries have been omitted, either to simplify the discussion by focusing on the principal country or countries within a region (such as China in eastern Asia) or, in some instances, because of missing data. Also omitted in order to simplify the discussion are the Pacific Islands of Micronesia; Antarctica; and, except in a few examples, central Asia. The following sections list all the countries included in the population and economic statistics and related projections.

Industrial Regions

North America. This region includes *Canada* and the *United States.* Many geographers would also include Mexico, and well within the fifty-year period considered here, the Mexican economy may become fully industrialized and closely integrated with those of the United States and Canada. But for the purposes of the scenarios in this book and the projections that support them, Mexico is included in Latin America.

Europe. This region comprises conventional western Europe—specifically, the countries of the European Union as well as nonmembers Norway and Switzerland. Within the fifty-year period addressed in this book, it is likely that some or perhaps all of the countries presently considered part of eastern Europe will become members of the European Union as the economic integration of Europe spreads eastward, but here eastern Europe is treated separately. In this book, the region of Europe includes *Austria, Belgium, Denmark, Finland, France, Germany, Greece, Ireland, Italy, the Netherlands, Norway, Portugal, Spain, Sweden, Switzerland,* and *the United Kingdom.*

Japan. Australia and New Zealand, the other highly developed countries in the western Pacific region, are omitted not because they are unimportant but in order to focus on Japan. Scenarios for Australia and New Zealand would necessarily be quite different from those for Japan, which has distinctive problems and cultural characteristics.

Transitional Regions

Transitional regions are those in transition from centrally planned economies and authoritarian Communist regimes to market economies and democratic regimes; hence the name.

Russia. All of the Russian Federation is included in this region, but all of the independent republics that were formerly part of the Soviet Union are omitted.

Eastern Europe. This region includes the seven largest countries of eastern and central Europe and omits a number of smaller countries. Included are *Belarus, Bulgaria, the Czech Republic, Hungary, Poland, Romania,* and *Ukraine.*

Developing Regions

Latin America. This region includes Mexico and all of Central America, South America, and the Caribbean and omits some small, mostly island, countries. Included are *Argentina, Bolivia, Brazil, Chile, Colombia, Costa Rica, Cuba, the Dominican Republic, Ecuador, El Salvador, Guatemala, Haiti, Honduras, Jamaica, Mexico, Nicaragua, Panama, Paraguay, Peru, Puerto Rico, Uruguay,* and *Venezuela.*

China. China is effectively a region unto itself. Not included are other eastern Asian countries such as Mongolia and North Korea. South Korea is included in the Southeast Asia region, with which it is economically linked.

Southeast Asia. This region includes eight countries and omits the thinly populated and relatively undeveloped countries of Laos and Cambodia, for which reliable data are scarce. Included are *Burma, Indonesia, Malaysia, Philippines, Singapore, South Korea, Thailand,* and *Vietnam.*

India and Southeast Asia. The scenarios focus on India, by far the largest and most critical country in South Asia. However, the region as a whole is about one-third larger in population than India alone, so statistics are also given for the six countries of *South Asia,* which include *Bangladesh, Bhutan, India, Nepal, Pakistan,* and *Sri Lanka.*

Sub-Saharan Africa. This region includes forty-three African countries south of the Sahara Desert, including the island countries of Madagascar and Mauritius. Included are *Angola, Benin, Botswana, Burkino Faso, Burundi, Cameroon, the Central African Republic, Chad, Côte d'Ivoire, the Democratic Republic of Congo* (formerly Zaire), *Equatorial Guinea, Eritrea, Ethiopia, Gabon, Gambia, Ghana, Guinea, Guinea-Bissau, Kenya, Lesotho, Liberia, Madagascar, Malawi, Mali, Mauritania, Mauritius, Mozambique, Namibia, Niger, Nigeria, the Republic of the Congo* (located to the east of Gabon and to the west of the Democratic Republic of Congo), *Rwanda, Senegal, Sierra Leone, Somalia, South Africa, Sudan, Swaziland, Tanzania, Togo, Uganda, Zambia,* and *Zimbabwe.*

North Africa and the Middle East. This region includes five North African countries and twelve Middle Eastern countries and omits a few smaller western Asian countries. Both Islam and oil link these two subregions together. North Africa includes *Algeria, Egypt, Libya, Morocco,* and *Tunisia;* the Middle East includes *Iran, Iraq, Israel, Jordan, Kuwait, Lebanon, Oman, Saudia Arabia, Syria, Turkey, the United Arab Emirates,* and *Yemen.*

Population Projections

The population projections used in this book are those released in 1997 by the United Nations Population Division and reflect a significant slowing of global population growth compared with earlier versions.[1] The following table gives the low, medium, and high U.N. projections,

Population Projections, 2050 (in millions)

	1995	2050		
		LOW	MEDIUM	HIGH
Industrial Regions				
North America	297	301	384	452
Europe	383	293	346	389
Japan	125	96	110	122
Transitional Regions				
Russia	148	97	114	142
Eastern Europe	152	110	131	156
Developing Regions				
Latin America	471	643	802	991
China	1,220	1,198	1,517	1,765
Southeast Asia	511	664	827	1,010
India[a]	929	1,231	1,533	1,885
Sub-Saharan Africa	586	1,518	1,783	2,089
North Africa and Middle East	348	650	785	930
WORLD TOTAL[b]	5,687	7,662	9,367	11,156

[a] The corresponding numbers for South Asia as a whole are 1,225 for 1995 and 1,786, 2,194, and 2,665 (low, medium, and high projections, respectively) for 2050.

[b] The regional numbers do not add up to the world total because a number of countries have been omitted to simplify the regional discussions.

together with the 1995 base population, for each of the regions described earlier.

Economic Projections

The gross national product (GNP), a measure of the economic output of a national economy, is an imperfect way to gauge economic activity. But it is widely used and most official statistics are based on it, so it was used in the analysis underlying this book. Similarly, per capita GNP is not exactly the same as average income for a country's citizens, but it is nonetheless widely used by the World Bank and other institutions as a measure of prosperity, a gauge of a country's economic development progress. In this book, per capita GNP is used in this way and is treated as a proxy for average income. That is, in discussion of a country's average income and comparison of average incomes within a region today or in the future, the numbers given represent per capita GNP.

To compare economic data for different countries, it is useful to convert local currencies into U.S. dollars, the international yardstick for money. There are two ways to do this. The older method is by using market exchange rates for different currencies. However, these not only fluctuate widely but also fail to measure accurately the real cost of living within a country. The newer method, called "purchasing power parity" (ppp), compares the cost of an equivalent basket of goods and services in different countries and calculates the correction factor required to make those goods and services cost the same in every country—the value of the local currency that gives equivalent purchasing power to that of the U.S. dollar in the United States. Thus average Japanese income (per capita GNP), measured at the market exchange rate for Japanese yen and U.S. dollars, appears to be higher than average U.S. income, but when the high cost of living in Japan is factored in—which is what the ppp system does—average Japanese incomes are seen to be lower than average U.S. incomes. Most economists believe that ppp figures, although not perfect, provide a more accurate measure of relative prosperity than do exchange rates.

Differences between average incomes (per capita GNP) calculated with exchange rates and those based on ppp figures are especially large for developing countries. The correction for equivalent purchasing power in China, for example, is a factor of five—that is, average Chinese incomes as measured by ppp are five times larger than when measured at current market rates for Chinese yuan, reflecting the very low cost of living in most of China. Another way to look at the ppp correction factor—one that is especially pertinent to the projections in this book—is to say that ppp figures anticipate a future time when a developing country's goods and services will be valued higher in international markets than they are now. In effect, ppp figures anticipate that exchange rates for Chinese currency—and for the currencies of most other developing countries—will rise over time; that the value of the currency will catch up with its internal purchasing power as assets such as productive land and unskilled labor rise in value (and hence as the prices paid for them also rise). Half a century from now, in theory, the gap between average incomes (per capita GNP) measured by purchasing power and average incomes measured with then current exchange rates should be much smaller for most developing countries, especially for countries that have export-oriented economies, such as China. Indeed, war-damaged Germany and Japan caught up with the United States in relative prosperity during postwar decades not just because of faster economic growth but also because the value of their increasingly skilled workforces—and hence of the goods and services they produced—rose sharply. Reflecting this revaluation, the exchange rate value of the German mark also rose, going from more than four to the dollar in the early 1960s to nearly two to the dollar in the early 1990s.

Thus, long-term projections done with conventional (exchange rate) GNP figures may significantly understate true prosperity, since prosperity depends not only on economic growth but also on the (potentially very large) change in value of the goods and services a country produces. On the other hand, long-term projections done with ppp figures may overstate future prosperity, since they assume that the value of a country's goods and services will rise—which might not happen, for example, in a country that isolates itself from the global economy or that completely fails to develop. On balance, the ppp approach seems better, more likely to give a useful set of long-term economic projections by region. Accordingly, all economic figures in this book are given in ppp terms unless otherwise noted.

The high and low economic projections are constructed to provide a consistent set of plausible limits for the economic output of different regions under the extreme assumptions, respectively, of very successful and sustained rapid growth over the next half century and of consistent failure to establish or sustain economic growth (possibly including extended periods of state failure or social chaos). The medium or conventional development projections assume continuity of current policies and practices and are based on—more accurately, adapted from—base case or "business-as-usual" scenarios of the Intergovernmental Panel on Climate Change, one of the more extensive recent efforts to project future economic growth.

For developing regions, the low economic projection assumes a 2 percent annual growth rate over the fifty-five-year period from 1995 to 2050. One way to express the cumulative effects of this pattern of growth is the growth factor—the multiplier by which the GNP in 2050 exceeds the GNP in 1995. For a 2 percent growth rate, the multiplier is 2.972, so the low projection assumes that economic activity in developing regions roughly triples. The medium economic projection assumes growth rates of 3.5 percent annually, equivalent to a growth factor of 6.633; and the high projection assumes growth rates of 4.5 percent annually, equivalent to a growth factor of 11.256.

Growth rates for transitional regions are assumed to be lower than for developing regions, and those for industrial regions lower still. For the industrial regions, for example, the high economic projection assumes a return to 1950s-style growth, about 3 percent per year, for several decades; the low projection would reflect near economic stagnation.

All projections start with World Bank figures for 1995 GNP[2] and are given in constant 1995 U.S. dollars; that is to say, they exclude inflation. For a few countries for which current GNP calculations are not available, estimated values of GNP have been used to make regional totals more realistic. The growth rate assumptions (and growth factors) for all regions are summarized in the first table that follows. The economic projections that result from these growth assumptions, together with the 1995 GNP base figures, are given in the second table.

Growth Rate Assumptions, 1996–2050

	LOW PROJECTION		MEDIUM PROJECTION		HIGH PROJECTION	
	Growth Rate	Growth Factor	Growth Rate	Growth Factor	Growth Rate	Growth Factor
Industrial regions	1.5/1%[a]	2.0	2.5/1.5%	3.044	3/2.5%	4.5
Transitional regions	2%	2.972	3%	5.082	4%	8.646
Developing regions	2%	2.972	3.5%	6.633	4.5%	11.256

[a]Where two numbers are given, the first represents the annual economic growth rate for the period 1996–2025; the second, the period 2026–2050.

Economic Projections, GNP (billions of constant 1995 U.S. dollars)

	1995	2050		
		LOW	MEDIUM	HIGH
Industrial Regions				
North America	7,828	15,656	23,828	35,226
Europe	7,337	14,674	22,334	33,017
Japan	2,765	5,530	8,417	12,443
Transitional Regions				
Russia	665	1,976	3,380	5,750
Eastern Europe	678	2,015	3,446	5,862
Developing Regions				
Latin America	2,677	7,956	17,757	30,132
China	3,563	10,589	23,633	40,105
Southeast Asia	2,229	6,625	14,785	25,090
India[a]	1,301	3,867	8,630	14,644
Sub-Saharan Africa	773	2,297	5,127	8,701
North Africa and Middle East	1,731	5,145	11,482	19,484

[a]The corresponding numbers for South Asia as a whole are $1,853 for 1995 and $5,507, $12,291, and $20,857 (low, medium, and high projections, respectively) for 2050.

Best- and Worst-Case Analyses

Economists usually assume that rising populations mean increased economic activity and higher economic growth, and in developed countries this generally has been the case. But economic growth and population growth are not directly coupled and, in some circumstances, can even vary inversely. Indeed, in developing countries, where there are large pools of surplus labor, population growth does not necessarily increase economic activity very much, and it can actually decrease average incomes (per capita GNP) where economic growth is slow, as in much of sub-Saharan Africa during the past few decades. Conversely, increased prosperity would be expected to reduce fertility and slow population growth.

In combining economic and demographic projections to estimate average incomes, it makes sense to look at the best-case and worst-case circumstances in each region. In developing regions, the best case arises when economic growth is high and population growth is low. Combining these projections gives the best-case average income (per capita GNP), the highest average prosperity. Correspondingly, the worst-case average income arises from combining the low economic growth projection with the high population growth projection. These are extremes, meant only to set outer limits of the plausible range of economic development over the next half century under very optimistic and very pessimistic scenarios.

For the industrial and transitional regions, where total economic activity is likely to be closely correlated with the size of the workforce, the best case arises when the high economic projection is combined with the high population projection. The worst case arises when the low economic projection is combined with the low population projection, yielding a narrower range of plausible average incomes. Midrange estimates of prosperity or average income arise from combining medium economic and population projections in all regions. The following table gives per capita GNP figures for 1995 and for 2050, corresponding to the projections in this book, which are used, as described earlier, as a proxy for average income or relative prosperity.

Plausible Range of Economic Development by 2050 (per capita GNP, constant 1995 U.S. dollars)

	1995	2050		
		WORST CASE	MIDRANGE CASE	BEST CASE
Industrial Regions				
North America	26,357	52,013	62,053	77,934
Europe	19,157	50,082	64,549	84,875
Japan	22,120	57,604	76,515	101,988
Transitional Regions				
Russia	4,493	20,375	29,645	40,490
Eastern Europe	4,461	18,153	26,302	37,577
Developing Regions				
Latin America	5,684	8,028	22,140	46,862
China	2,920	6,000	15,579	33,477
Southeast Asia	4,362	6,559	17,878	37,786
India[a]	1,400	2,051	5,629	11,896
Sub-Saharan Africa	1,319	1,100	2,876	5,732
North Africa and Middle East	4,974	5,532	14,626	29,976

[a]The corresponding numbers for South Asia are $1,513 for 1995 and for 2050, $2,066, $5,602, and $11,678 (worst, midrange, and best cases, respectively) for 2050.

Environmental Projections

The environmental projections in this book are adapted from detailed projections contained in the Conventional Development Scenario (CDS) prepared by the Stockholm Environment Institute.[3] The CDS projections assume moderate economic growth: they are based on base case, or "business-as-usual," assumptions of economic growth rates, roughly equivalent to the midrange economic projections in this book. However, the CDS projections include the structures of regional economies by sector as well as detailed projections of energy demand for each region. The CDS projections were used to estimate the potential growth of air pollution and toxic emissions.

Energy production and use are major sources of pollution, particularly air pollution. So, as developing economies industrialize and increase their use of energy, air pollution is also likely to increase. The primary contributors to more long-lived toxic pollution are the metals and chemicals sectors of industrial economies and, to a lesser degree, pulp and paper production. A number of studies suggest that as developing economies industrialize, these sectors are likely to increase their output—and hence the potential for toxic emissions—roughly in proportion to the growth of the industrial sector of the economy. Many of these toxic materials are dumped into rivers or ultimately find their way into streams and estuaries, making them a primary cause of hazardous water pollution.

Overall, the CDS projections suggest that energy use worldwide will grow by a factor of more than two and one-half over the next half century, even taking into account expected improvements in efficiency and assuming rapid transfer of more efficient practices to developing economies. The CDS projections also suggest that worldwide industrial activity may increase by a factor of three. In rapidly industrializing regions, however, the growth factors are much larger: China's energy consumption is projected to grow sixfold and its industrial activity more than tenfold.

These projected increases do not necessarily mean that air pollution in China will increase by a factor of six or that the region's toxic emissions will increase more than tenfold. But they do suggest the size of the increase in polluting activities and hence the potential for increased pollution if current industrial patterns and practices continue.

The projected growth in energy use—an energy growth factor that also gauges the potential increase in air pollution—for each region is given in the table that follows, based on the CDS projections adjusted for the period 1995–2050. The table also gives the projected growth of

Air and Toxic Pollution Projections, 1995–2050

	Air Pollution Factor (Energy Growth Factor)	Water Pollution Factor (Industrial Growth Factor)
Latin America	2.5	6.4
China	6.0	10.6
Southeast Asia	5.2	8.6
India/South Asia	5.2	9.8
Sub-Saharan Africa	5.3	8.5
North Africa and Middle East	4.9	6.3

the industrial sector of each region's economy—an industrial growth factor that also gauges the potential increase in toxic emissions and water pollution.

Potential Regional Effects of Climate Change

Even though the Conventional Development Scenario suggests that worldwide energy use will grow by more than two and one-half over the next half century, it might, of course, grow even more under conditions of high economic growth. Under prevailing patterns and policies, nearly all that energy would come from fossil fuels—coal, oil, and natural gas. That means emissions of carbon dioxide—the greenhouse gas released during combustion of these fuels—would also be expected to increase by about two and a half times over the next half century. Under those conditions, "greenhouse" warming of the atmosphere would be expected to accelerate.

Scientists cannot yet say with certainty just how severe an effect more greenhouse warming would have on the climate or how soon the changes might be apparent. Regional effects are still less certain. Nonetheless, a recent study by an international team of scientists gives their best judgment—based on modeling efforts and historical studies— of the likely regional consequences.[4] They include warmer temperatures; greater variability in weather and more extreme weather, such as heat waves; both increases and decreases in precipitation; longer drought cycles; more flooding; and a gradual increase in sea level. These in turn are expected to cause crop failures; significant changes in forest cover and rangelands, including increased desertification, changes in water supplies, spread of disease vectors leading to potential effects on human health; and large-scale migration from inundated coastal and floodplain areas. The following is a brief summary of some of the highlights of the international study.

In Latin America, rangelands and the livestock operations that depend on them may be the most severely affected. Northern China may see reduced precipitation, worsening its water shortages, whereas southern and coastal China may be more vulnerable to increased flooding. Inundation of low-lying areas could displace millions of people in

South Asia, which could also have adverse effects on human health and on agriculture, especially among traditional farmers. The most severe consequences for humans may be experienced in sub-Saharan Africa, where thirty-six countries are already affected by drought and desertification, both of which are expected to increase. North Africa and the Middle East may see some increased rainfall, but higher temperatures are likely to mean reduced soil moisture and more desertification, harming agriculture. Runoff could decrease significantly in some areas, exacerbating already severe water shortages. In the industrial and transitional regions, North America may see a northward shift of forest cover, and less runoff and reduced water supplies in the Midwest, and Europe may find it wetter in the north but drier in the south and east. The greatest temperature changes are expected in high latitudes, extending the growing season in Canada and northern Russia and melting some of the permafrost.

Potential Scarcity of Land and Water

A recent U.N. study projects that withdrawals of freshwater for human use will increase by 40 percent in the first twenty-five years of the new century.[5] Based on this rising water use, the study projects that by the year 2025, virtually all of North Africa and the Middle East, all of South Asia, and southern Africa will face chronic water scarcity; China and parts of North America and Europe will face less severe shortages; and water problems will begin to appear in parts of Southeast Asia, East Africa, and the Caribbean. In all, more than two-thirds of the world's population may well experience "moderate to high" difficulties with water management by 2025, and these problems are expected to worsen by 2050, suggesting that water scarcity will become a critical factor in regional development.

Potential scarcity of water and of farmland can be estimated in another way, by comparing the renewable water supply or the amount of arable land for each country, assumed to remain constant, with projected populations and seeing whether the resulting ratios fall beneath critical values.[6] This method suggests that by the year 2050, 1 billion to 2.4 billion people will be living in water-scarce countries and 1.6 billion to 5.5 billion people will be living in land-scarce countries. (The range of numbers corresponds to the high and low U.N. population projections.) Land scarcity will affect China, India, and many countries in sub-Saharan Africa.

These projections do not necessarily mean that countries within these regions will face actual scarcity—land can be farmed more intensively, and water can be cleaned up and reused. But to do so usually requires financial investments and technical knowledge beyond the capacity of traditional farmers in poor rural areas. So these potential scarcities could become a real constraint in the most vulnerable rural communities.

Notes

Preface

1. Murray Gell-Mann, *The Quark and the Jaguar* (New York: W. H. Free-man, 1994), 346. This insightful volume provides clear descriptions and many examples of complex systems.

2. The *World Resources* series is prepared by the World Resources Institute in collaboration with the United Nations Environment Programme, the United Nations Development Programme, and the World Bank. Volumes include World Resources 1990–91, 1992–93, 1994–95, 1996–97, 1998–99, published by Oxford University Press, New York, in 1990, 1992, 1994, 1996, and 1998, respectively. I served as editor-in-chief of this 400-page reference work for three editions, with continuing administrative oversight for subsequent volumes.

3. The international data on economic and social conditions used in this book are contained in the annual editions of the *World Development Report* series prepared by the World Bank and the *Human Development Report* series prepared by the United Nations Development Programme (both series published by Oxford University Press, New York).

Chapter 1 Thinking About the Future

1. The thoughts in this paragraph appeared in an earlier form in Gilberto Gallopin, Allen Hammond, Paul Raskin, and Rob Swart, *Branch Points: Global Scenarios and Human Choice* (Stockholm: Stockholm Environment Institute, 1997).

Chapter 2 The Power of Scenarios

1. Peter Schwartz, *The Art of the Long View* (New York: Doubleday, 1991).

2. Ibid., 7–9.

3. Business Council for Sustainable Development, in press (Geneva: Geneva World Business Council for Sustainable Development, 1998).

4. Ged R. Davis, "Energy for Planet Earth," *Scientific American* 263, no. 3 (1990): 57.

5. *The Evolution of the World's Energy Systems* (London: Shell International Ltd., 1996).

6. J. T. Houghton, L. G. Meira Filho, B. A. Callender, N. Harris, A. Kattenberg, and K. Maskell, eds., *Climate Change 1995: The Science of Climate Change*, Published for the Intergovernmental Panel on Climate Change, in collaboration with the World Meteorological Organization and the United Nations Environment Programme (Cambridge: Cambridge University Press, 1996).

7. Dirk Bryant et al., *Forest Frontiers* (Washington, D.C.: World Resources Institute, 1996); Dirk Bryant et al., *Reefs at Risk* (Washington, D.C.: World Resources Institute, 1998).

8. Robert Kaplan, "The Coming Anarchy," *Atlantic Monthly* (February 1994). Kaplan's somewhat sensational article identified a real concern but gave only a superficial view of the causes of African instability.

Part II Introductory Text

1. I have renamed these scenarios for this book, so they differ from those developed by the Global Scenario Group; in fact, the Global Scenario Group identified a total of six scenarios, two in each world, so the set of scenarios I present lacks the full complexity and richness of the original work. Interested readers are urged to consult the initial report of the Global Scenario Group. Gilberto Gallopin, Allen Hammond, Paul Raskin, and Rob Swart, *Branch Points: Global Scenarios and Human Choice* (Stockholm: Stockholm Environment Institute, 1997).

2. Ibid., 11

Chapter 3 Market World: *A New Golden Age of Prosperity?*

1. Richard Rosecrance, "The Rise of the Virtual State," *Foreign Affairs* 75, no. 4 (1996): 45.

2. This scenario takes its name and some of its attributes from a similar scenario described by Peter Schwartz in *The Art of the Long View* (New York: Doubleday, 1991); it also draws on the Conventional Worlds scenarios of the Global Scenario Group, as described in Gilberto Gallopin, Allen Hammond, Paul Raskin, and Rob Swart, *Branch Points: Global Scenarios and Human*

Choice (Stockholm: Stockholm Environment Institute, 1997), and on a more recent scenario by Schwartz and Peter Lyden published in *Wired* magazine, "The Long Boom," *Wired* (July 1997): 115–173.

3. The phrase *Asian tigers* initially referred to South Korea, Taiwan, Singapore, and Hong Kong to describe their rapid economic growth in the 1970s and 1980s.

4. G. Paul Zachary, "Global Growth Attains a New, Higher Level That Could Be Lasting," *Wall Street Journal*, 13 March 1997, A1.

5. World Bank, *Global Economic Prospects and the Developing Countries* (Washington, D.C.: World Bank, 1995), 18.

6. Erik R. Peterson, "Surrendering to Markets," *Washington Quarterly* 18, no. 4 (1996): 103–115.

7. Jesse Ausubel, "Can Technology Spare the Earth?" *American Scientist* 84 (March–April 1996): 166–178.

8. William H. Gates, "Engaging Science—Sustaining Society" (paper presented at the annual meeting of the American Association for the Advancement of Science, Seattle, Wash., 17 February 1997). For the year 2011, Gates forecast computer chips with 1 billion transistors (compared with 5.5 million in 1997), operating at 10 billion cycles per second and capable of performing 100 billion instructions per second.

9. Despite higher energy efficiency, for example, global energy consumption is expected to rise sharply as population and economic growth outrun efficiency improvements. And continuing improvements in agricultural yields depend on continued research and on the spread and adoption of improved practices, which in turn may depend on continued growth in prosperity and on stable societies. Also important are whether water is available for irrigation, whether the fertility of existing soils is degraded, whether farmers are educated enough to make use of improved techniques, and a host of other factors. Agricultural researchers in particular are concerned that gains in yields are slowing and that international investment in agricultural research is declining.

10. The three historical waves or clusters of innovation in the United States correspond to periods of rapid economic growth in the early 1800s, during the forty years prior to World War I, and in the post–World War II period from 1950 to 1973. Arnulf Grubler, "Time for a Change: On the Patterns of Diffusion of Innovation," *Daedalus* 125, no. 3 (1996): 19–42.

11. Zachary, "Global Growth."

12. It can take a long time—decades—for radically new technologies to progress incrementally, penetrate into society, and gain widespread acceptance. The antecedents for the information revolution have been edging their way into society for some time now. The first crude computers were built during World War II. The transistor emerged in 1950; the first integrated circuits, or chips, were made about 1960, as was the first laser. The first optical fibers were man-

ufactured in the late 1960s; the first commercial microprocessor (which made personal computers possible), in 1970. Yet it was not until the 1980s that personal computers appeared; not until the 1990s that lasers and optical fibers had improved enough to begin carrying most U.S. long-distance telephone traffic and personal computers had become so inexpensive and ubiquitous that linking them together in large networks became interesting. It was not until the foregoing infrastructure was in place that the development of on-line commerce, extensive electronic mail networks, and on-line communities based on common interests could begin.

13. The steam engine, harnessing energy to drive machines, was an enabling technology that increased by a factor of about 1,000 the power that could be applied to a given task—a change from one human or one horsepower to a steam engine capable of exerting 1,000 horsepower. It took a century and the invention of the automobile before ordinary individuals, rather than industrial companies, had that kind of power available to them.

For computing power and telecommunications bandwidth, the comparable enabling technologies of the information revolution, the period from 1980 to 2010 will see an increase in the computing power available in a personal computer by a factor of 100,000 and an increase in the bandwidth of a single optical fiber by a factor of at least 10,000. By these measures, then, the information revolution represents a far larger technological shift than did the industrial revolution. The speed of technological progress continues to outpace official estimates. In the late 1980s, for example, the U.S. government set some end-of-the-century goals for demonstrating high-speed computation and communication technology: a teraflop computer and terabit optical communications. A teraflop computer would complete 1 trillion operations per second; a terabit optical fiber would transmit 1 trillion bits of information per second.

In 1996, however, a Japanese computer company announced that its new supercomputer could calculate at teraflop speeds, working a billion times faster than early personal computers; Intel Corporation, an American company, announced that its new "massively parallel" supercomputer had also achieved those speeds; and U.S. government scientists at Sandia National Laboratories were already planning a computer 1,000 times as fast. Moreover, communications experiments carried out by three different firms—Fujitsu Ltd., Nippon Telegraph and Telephone Corporation, and AT&T Corporation—achieved experimental transmission of information at terabit rates. That rate was some 400 times faster than the rate of then current commercial systems and is the equivalent of transmitting 12 million telephone conversations simultaneously through a single fiber or transmitting the contents of the Library of Congress in the blink of an eye. Now the U.S. government is aiming at computers that can operate at 100 teraflops.

Such awesome computational and communication power may not routinely be needed at the beginning of the new century, but as demand for ever faster computation and communications rises, these advances will begin to be incorporated in specialized and leading-edge applications. Moreover, even commonplace personal computers in the year 2001 will be likely to operate at a "clock rate" of 1 billion cycles per second, and many long-distance communications lines will carry information at the rate of 50 billion bits per second.

That is to say, the information infrastructure that we use every day and largely take for granted is already close to operating at a billion bits per second, with trillion-bit-per-second capability or more waiting in the wings. It is our minds, our organizations, and our social infrastructure that are not yet ready for such speeds.

14. The information revolution is making possible, for example, the following: An engineer at the Boeing Company can gain access from his workstation to complete plans and manufacturing specifications for each of the millions of parts of a new 757 airliner at each stage of its design. A bond trader in London or Tokyo can instantly read about and bid on a new offering at the same time as do her counterparts in Chicago and New York. A small team at the headquarters of Wal-Mart Stores, Inc., can daily review the previous day's sales figures for each of 10,000 items at stores all over the world and adjust pricing and replacement orders accordingly. University scientists in Michigan and Norway can jointly monitor and discuss data being remotely collected by a complex instrument on the ice cap in northern Greenland while directing the instrument's activities from their campus offices.

15. Such widespread effects would not be surprising. The industrial revolution, after all, transformed the economic and social patterns of eighteenth-century England and nineteenth-century Europe and the United States. The economic and social consequences of the information revolution are likely to be greater, even though they may not fully appear for several decades.

16. Beginning in 1997, several competing international consortia, such as the Iridium consortium headed by Motorola, are launching satellite-based communications systems that rely on large numbers of satellites circling in low orbits, close enough that special cellular phones can communicate directly with them. The satellites will be interconnected with ground-based telephone systems. People using the systems will be able to make and receive calls (and, later, send and receive data) from anyplace on Earth—from a major metropolitan area, from a village in a jungle, from a boat at sea. The implications of these systems are discussed further in chapter 5.

17. Lester Thurow, "The Revolution upon Us," *Atlantic Monthly* (March 1997): 97–100.

18. Ervin Laszlo, "Dimensions of a New World Order," in *Visions for the Twenty-First Century,* ed. Sheila M. Moorcroft (Westport, Conn.: Praeger, 1993), 9–17.

Chapter 4   Fortress World: *Instability and Violence?*

1. Madhav Gadgil and Ramachandra Guha, *Ecology and Equity: The Use and Abuse of Nature in Contemporary India* (London: Routledge, 1995), 34.

2. This scenario draws heavily on the work of the Global Scenario Group (GSG) and in particular on the insights of my GSG colleagues Paul Raskin and Gilberto Gallopin. The name *Fortress World* comes from one of the GSG scenarios, but I use it here in a broader sense. See Gilberto Gallopin, Allen Hammond, Paul Raskin, and Rob Swart, *Branch Points: Global Scenarios and Human Choice* (Stockholm: Stockholm Environment Institute, 1997), especially the family of Barbarization scenarios.

3. A full answer requires the detailed analysis of current trends provided in chapters 6–9 and the consideration of specific regional contexts given in chapters 10–16.

4. United Nations Development Programme, *Human Development Report* (New York: Oxford University Press, 1996), 1.

5. Thomas Friedman, "Yesterday's Man," *New York Times,* 19 March 1995, E15.

6. "Time to Roll Out a New Model," *Economist,* 1 March 1997, 71–72.

7. M. S. Swaminathan, personal communication, 1993.

8. Gadgil and Guha, *Ecology and Equity.*

9. Peter H. Gleick, ed., *Water in Crisis: A Guide to the World's Freshwater Resources* (New York: Oxford University Press, 1993), 276.

10. Matthew Connelly and Paul Kennedy, "Must It Be the Rest Against the West?" *Atlantic Monthly* (December 1994); World Resources Institute, *World Resources 1994–95* (New York: Oxford University Press, 1994), 31.

11. See the discussion in chapter 6 for documentation of equity trends.

12. "Welcome to the New World of Private Security," *Economist,* 19 April 1997, 21–24.

13. A. M. Rosenthal, "Only a Matter of Time," *New York Times,* 22 September 1996, A31.

Chapter 5   Transformed World: *Changing the Human Endeavor?*

1. This quote is widely attributed to Margaret Mead but is not found, as far as I can establish, in her published writings.

2. In virtually every country, governments used markets to ration energy use rather than creating bureaucracies, either by raising energy taxes or by auctioning off energy use permits that could be traded. Likewise, the legislation

setting up energy rationing in the United States, and that in most other industrial countries, required that tax or permit revenues be used to reduce social security or other employment taxes, recycling the money into the economy— what economists call a "revenue-neutral" approach—to minimize the economic pain of rationing. The emissions trading scheme proposed by President Clinton is a step toward such a system.

3. For this description of the growing role of citizen's groups and some of the other ideas in this scenario, I am indebted to my former colleague at World Resources Institute Jessica Mathews and her farseeing article "Power Shift," *Foreign Affairs* 76, no. 1 (1997): 50–66.

4. Bruce Smart, ed., *Beyond Compliance: A New Industry View of the Environment* (Washington, D.C.: World Resources Institute, 1992).

5. Stephan Schmidheiny with the Business Council for Sustainable Development, *Changing Course: A Global Business Perspective on Development and the Environment* (Cambridge, Mass.: MIT Press, 1992). Also Paul Hart, "Beyond Greening: Strategies for a Sustainable World," *Harvard Business Review*, January–February 1977.

6. Joe Klein, "In God They Trust," *New Yorker*, 16 June 1997, 39–48. As Klein points out, studies of similar efforts suggest that church-based efforts are more successful than other social programs at preventing juvenile crime, reclaiming drug addicts and alcoholics, and educating inner-city kids.

7. See, for example, Steve Lerner, *Eco-Pioneers* (Cambridge, Mass.: MIT Press, 1997).

8. Jake Page, "Ranchers Form a "Radical Center" to Protect Wide Open Spaces," *Smithsonian* 28, no. 3 (1997): 50–56.

9. Jessica Mathews, "The New, Private Order," *Washington Post*, 21 January 1997, A1.

10. Robert H. Anderson et al., *Universal Access to E-Mail* (Santa Monica, Calif.: Rand Corporation, 1995), 126–127.

11. Esther Dyson, "The Power of the Internet," *Washington Post*, 19 January 1997, C4.

12. Youssef M. Ibrahim, "All the Sights of the City Just a Mouse Click Away," *New York Times*, 11 July 1997.

13. United Nations Development Programme, *Human Development Report 1997* (New York: Oxford University Press, 1997).

Part III   Introductory Text

1. In 1960, South Korea was one of the world's poorest nations, with an average annual per capita income of about $360, few natural resources, and a society still recovering from the ravages of the Korean War. Yet over the next thirty-five years, Korea transformed itself into a modern industrial nation.

Income rose by more than 6 percent per year, doubling more than three times and bringing a standard of living nearly comparable to those of some European nations—despite a population density that today is nearly four times that of Nigeria. Korea's successful strategy, starting with an emphasis on education and pursuing first land reform and rural development, then export-led industrialization, became a textbook case of good national economic management—a success not diminished by its recent failure to manage its banking system properly.

Nigeria started with similar economic circumstances in 1960 but discovered a vast pool of oil in what should have been a stroke of good fortune. Economic growth accelerated, and Nigeria is today a major oil exporter. But a series of corrupt military governments squandered or stole most of the oil wealth; thus, very little has reached ordinary citizens. Per capita income grew very slowly and in the past fifteen years has dropped precipitously, leaving the average Nigerian worse off economically than before oil was discovered and with a climate of repression and human rights violations that has made its government an international pariah.

2. If we take India's present annual per capita gross national product of $1,400 as an approximation of average annual income, then growth of the country's economy at 4.5 percent per year between 1995 and 2050 would boost this figure to about $12,000, less than half current average U.S. income, and then only if India's population grows more slowly than expected.

3. Analysis of critical trends is most useful for trends that persist over some period of time, usually decades or more. A persistent trend usually means that the underlying phenomenon changes relatively slowly. Examples include fertility, the average number of children that women in a given society will have; and agricultural yields, how much of a crop the average farmer can grow per acre. Because fertility and agricultural yields tend to change only slowly—unless societies undergo drastic social and economic upheaval—projecting or modeling future trends in population size and agricultural production is relatively easy and reliable. Even technology, which today seems to evolve with lightning speed, often follows a fairly consistent pattern in the rate of improvements in a given field.

Chapter 6   Critical Trends: Demographic, Economic, and Technological

1. For technical reasons, the replacement rate fertility for a population is about 2.1 children per woman, averaged over the population, rather than exactly 2 children.

2. Santiago is the capital of Chile, the nation with the most rapidly growing economy in Latin America and whose privatized social security system is

already attracting worldwide attention. Bangalore, in southern India, is already known as a hotbed of high-tech activity, the emerging Silicon Valley of that country.

3. The International Food Policy Research Institute (IFPRI) has published an extensive series of documents under its 2020 Vision project that support this basic conclusion. See, for example, Mark Rosegrant et al., *Global Food Projections to 2020: Implications for Investment* (Washington, D.C.: International Food Policy Research Institute, 1995).

4. Per Pinstrup-Anderson, Rajul Pandya-Lorch, and Mark Rosegrant, *The World Food Situation: Recent Developments, Emerging Issues, and Long-Term Prospects* (Washington, D.C.: International Food Policy Institute, 1997), 8.

5. The global economy expanded by more than 3.5 percent over the 1994–1997 period, much more rapidly than in the preceding two decades. Some economic forecasters, such as the WEFA Group in Eddystone, Pennsylvania, expect global growth to average as much as 4 percent over the next two decades.

6. World Bank, *Global Economic Prospects and Developing Countries* (Washington, D.C.: World Bank, 1997). The report forecasts not only that growth will accelerate in developing countries but also that China, India, Indonesia, Brazil, and Russia will become economic powerhouses in the next quarter century, doubling their share of world output.

7. This is not as implausible as it may sound for developing regions where the conditions for rapid economic growth have been established. Until 1997, South Korea's economy grew by more than 6 percent annually for more than thirty-five years, nearly doubling the country's income every decade.

8. In both India, which already has a middle class of some 200 million people, and sub-Saharan Africa, where there is wide variation among the forty-four countries in the region, there will be significant subpopulations that can attain much higher incomes. Nonetheless, the average figures given are an important indicator of the prosperity of a region as a whole.

9. World Resources Institute, *World Resources 1994–95* (New York: Oxford University Press, 1994), 5. These gaps in average per capita GDP are measured in conventional (exchange rate) terms, unlike the purchasing power parity figures used elsewhere in the book.

10. Economists usually compare the gap between rich and poor countries in relative terms—average income in one country as a percent of average income in another—rather than measuring the absolute differences in dollars, as reported here. Most people, however—presumably including potential migrants—compare incomes in dollars, not percentages. But even in relative or percentage terms the gap between incomes in developing and industrial countries is narrowing for only about a dozen developing countries, although these include some of the largest, such as China, India, and Brazil. "Over the next

generation," one recent study concluded, "if current trends continue, some 10–12 Southern nations are likely to join the ranks of the North or at least move much closer to Northern levels of economic performance. The remaining 140-odd Southern nations, however, are likely to slip further behind." Robin Broad and Christina Melhorn Landi, "Whither the North–South Gap?" *Third World Quarterly* 17, no. 1 (1996): 7–17.

11. United Nations Development Programme, *Human Development Report 1992* (New York: Oxford University Press, 1992), 34.

12. United Nations Development Programme, *Human Development Report 1996* (New York: Oxford University Press, 1996), 170–171.

13. Patrick E. Tyler, "On the Farms, China Could Be Sowing Disaster," *New York Times,* 10 April 1995.

14. The top 5 percent of American households received 11.5 times as much income as the bottom 20 percent in 1970; by 1990, the ratio had increased to 15—mean incomes of $148,124 and $9,833, respectively. See U.S. Department of Commerce, Bureau of the Census, *Current Population Reports,* Series P-650, no. 174 (Washington, D.C.: U.S. Government Printing Office, 1991); see also World Resources Institute, *World Resources 1994–95,* 19. According to Edward Wolff of New York University, author of a recent study on equity trends, America is "the most unequal industrial country in terms of income and wealth, and we're growing more unequal faster than the other industrial countries," *New York Times,* 17 April 1995, 17. See also Edward N. Wolff, *Top Heavy: A Study of the Increasing Inequality of Wealth in America* (New York: Twentieth Century Fund Press, 1995).

15. Organization for Economic Cooperation and Development, *Technology and Industrial Performance* (Paris: Organization for Economic Cooperation and Development, 1997).

16. Advanced gas turbines that are cheaper to build than coal-burning or nuclear power plants and that generate electricity far more efficiently and with less pollution exist now, and some U.S. utilities are starting to install them. But since demand for electricity is growing only slowly and utilities generally do not plan to shut down older plants until they reach the end of their life, it could be several decades before the new plants come into widespread use in the United States. And lack of natural gas or the pipelines to carry it is limiting use of gas turbines in rapidly industrializing regions.

17. Richard J. Barnet, "The End of Jobs," *Harper's Magazine* (September 1993): 47–52.

18. Robert U. Ayres, *Turning Point: An End to the Growth Paradigm* (London: Earthscan, 1998).

Chapter 7   Critical Environmental Trends

1. These services include nutrient recycling, water purification, flood prevention, oxygen production, and more than a dozen other essential life-supporting activities. Many are irreplaceable, but by one estimate they are worth $33 trillion per year—more than the current output of all the world's economies. Robert Costanza et al., "The Value of the World's Ecosystem Services and Natural Capital," *Nature* 387, no. 6630 (1997): 253–260.

2. In 1995, the industrial countries emitted more than 40 percent of the carbon dioxide that is the principal greenhouse gas arising from industrial activity. Cumulatively, however, these countries are the source of more than 80 percent of carbon dioxide emissions since 1950—and since carbon dioxide molecules stay in the atmosphere for about 100 years on average, it is the cumulative emissions that have an effect on climate.

3. Based on reporting by Thomas Friedman, *New York Times*, 20 March 1996, A19.

4. Ibid.

5. Bangkok is not the only polluted place. Of some twenty megacities around the world studied by the United Nations, all exceeded at least one World Health Organization standard, fourteen exceeded two, and seven exceeded at least three. United Nations Environment Programme and World Health Organization, *Urban Air Pollution in the Megacities of the World* (Oxford, England: Blackwell Reference, 1992).

6. Albert Adriaanse et al., *Resource Flows: The Material Basis of Industrial Economies* (Washington, D.C.: World Resources Institute; Wuppertal, Germany: Wuppertal Institute; The Hague: Netherlands Ministry of Housing, Spatial Planning, and Environment; Tsukuba, Japan: National Institute for Environmental Studies, 1997).

7. Ibid. Japan's total is lower because it uses less energy per person and its rice-based agriculture does not create much soil erosion.

8. Andrew Pollack, *New York Times*, 6 June 1996, D1.

9. World Resources Institute, *World Resources 1990–92* (New York: Oxford University Press, 1990), 148.

10. The environmental projections in this book are estimates of what may happen if industrial development follows current patterns, assuming the midrange economic projections and the medium population projections described in chapter 6. As described in the appendix, the environmental projections are based on detailed sector-by-sector and region-by-region studies of energy demand and industrial development done by the Stockholm Environment Institute to estimate the consequences of continuing present developmental patterns.

11. Economic projections can perhaps suggest how likely it is that each

region will eventually launch effective cleanup efforts. Economic studies based largely on the Asian tigers suggest a pattern: as a society develops, air and water pollution tends to increase until per capita incomes reach a certain level—in effect, until a society is rich enough to have the luxury of worrying about environmental issues. Then, pressures for government action increase and pollution levels start to drop. See World Resources Institute, *World Resources 1996-97* (New York: Oxford University Press, 1996), 162-164. Whether this theory is correct is hotly debated, but there is some evidence that effective environmental cleanup efforts begin when countries reach average incomes in the $3,000-$9,000 range. Thus, local pollution might begin to abate well before the middle of the twenty-first century in China, most of Southeast Asia, and Latin America but only later in India and most of sub-Saharan Africa—unless leaders in India and other less wealthy societies make different choices about their priorities from those who have gone before.

12. World Bank, *China 2020* (Washington, D.C.: World Bank, 1997).

13. John Briscoe, "When the Cup Is Half Full: Improving Water and Sanitation Services in the Developing World," *Environment* 35, no. 4 (1993): 10.

14. Raskin and Margolis, *Global Energy*, 32. See also World Resources Institute, *World Resources 1996-97*, 280-281.

15. One measure of how hard it will be to prevent climate change is the finding by an international scientific study that a 60 percent cut in current amounts of global emissions would be required to stabilize concentrations in the atmosphere at present levels. Still larger cuts will be needed if emissions climb as projected.

16. Intergovernmental Panel on Climate Change, *IPCC Synthesis Report* (Geneva: World Meteorological Organization and United Nations Environment Programme, 1995).

17. Thomas Karl, Neville Nicholls, and Jonathan Gregory, "The Coming Climate," *Scientific American* 276, no. 5 (1997): 78-83.

18. William Cline, in his book *The Economics of Global Warming* (Washington, D.C.: Institute for International Economics, 1992), estimates that the costs of a 10°C rise in temperature—higher than expected during the twenty-first century—might reach 6 percent of GDP.

19. An analysis of direct and indirect consumption patterns in India showed that the rural population that accounts for the bottom 50 percent of the income distribution spends almost all its available income on food (and, to a much lesser extent, on clothing), with virtually nothing spent on durable goods, energy, or transportation. This population is thus dependent on locally gathered natural resources to meet all other basic needs—including water, food, fodder for cattle, fuelwood, housing materials, and tools. *World Resources 1994-95* (New York: Oxford University Press, 1994), 1922.

20. Madhav Gadgil and Ramachandra Guha, *Ecology and Equity: The Use and Abuse of Nature in Contemporary India* (London: Routledge, 1995).

21. World Resources Institute, *World Resources 1992–93* (New York: Oxford University Press, 1992), 111–117.

22. Bruce Stutz, "The Landscape of Hunger," *Audubon* 95, no. 2 (1993): 54.

23. Ibid.

24. World Resources Institute, *World Resources 1996–97*, 203.

25. Dirk Bryant et al., *Frontier Forests* (Washington, D.C.: World Resources Institute, 1997). Only 40 percent of the remaining forests are unaltered and intact ecosystems, and 40 percent of those are threatened. Much of the remaining virgin forests are in the arctic regions of Canada and Russia and in the Amazon forest of South America; only small and increasingly threatened patches of intact natural forest remain in the rest of the world. And although tree plantations and other commercial forests are increasing in number, they do not provide the genetic diversity nor harbor as wide a range of other species as do natural forests.

26. Barbara Crossette, "Report Blames Poor Farmers for Depleting World's Forests," *New York Times*, 19 August 1996, A9.

27. World Resources Institute, *World Resources 1990–91* (New York: Oxford University Press, 1990), 176.

28. Partha Dasgupta, *An Inquiry into Well-Being and Destitution* (Oxford, England: Clarendon Press, 1993). Recent findings also suggest that children become economic liabilities rather than assets to their parents only when they are required to go to school and thus are no longer as available to help out at home. When parents must pay school fees, a particular burden to poor rural people with limited access to cash income, the effect is even stronger. See W. Penn Handwerker, *Culture and Reproduction: An Anthropological Critique of Demographic Transition Theory* (Boulder, Colo.: Westview Press, 1986), 3. Joel Cohen, in his book *How Many People Can the Earth Support?* (New York: Norton, 1995), argues that hard times—such as sharp economic contractions and, possibly, famine and civil war—can also cause couples to limit their fertility. But as long as children are a valuable asset, there is no incentive for poor families to limit their fertility. Indeed, fertility is generally highest in those countries (and within countries, generally those provinces) where rural poverty is greatest and also where female literacy is lowest—usually, but not always, the same places.

29. Myron Weiner, *The Child and the State in India* (New Delhi: Oxford University Press, 1991), 1, 10–11.

30. Scarcity of renewable resources does not necessarily lead to higher prices and more efficient use, as simplistic economic reasoning might suggest. Markets for such resources, when they exist at all, are often imperfect and ownership of the resource itself is often unclear. Some resources, such as grazing

lands, may be owned in common by an entire village. Other resources, such as forestlands, are often owned by governments, with local use rights established by tradition but rarely codified in law. Water rights are governed by a vast array of different customs and laws, many of which are often usurped when a dam is built upstream. Lack of ownership can affect both the day-to-day stewardship of a resource and normal economic responses (such as higher prices) to scarcity.

31. Vaclav Smil, *Global Ecology: Environmental Change and Social Flexibility* (London: Routledge, 1993); see also Robert Engleman and Pamela LeRoy, *Conserving Land: Population and Sustainable Food Production* (Washington, D.C.: Population Action International, 1995).

32. The number will depend on how fast population grows; the estimates correspond to the low and high U.N. population projections, the plausible range for each region described in chapter 6.

33. *Report of the Secretary-General on a Comprehensive Assessment of the Freshwater Resources of the World* (New York: United Nations, 1997).

34. Malin Falkenmark and Carl Widstrand, "Population and Water Resources: A Delicate Balance," in *Population Bulletin* (Washington, D.C., Population Reference Bureau, 1992), 19; Robert Engelman and Pamela LeRoy, *Sustaining Water: Population and the Future of Renewable Water Supplies* (Washington, D.C.: Population Action International, 1993), 18–22. An alternative method, projecting potential scarcity when water use exceeds 40 percent of available supplies, gives very similar results.

35. In these projections, the size of the resource—renewable supplies of water—is assumed to remain roughly constant in each country; a range of population assumptions is used, varying from the U.N. low-growth to high-growth projections, as discussed in chapter 6.

36. Use of crop wastes and cattle dung as fuels, however, deprives soils of needed nutrients and thus does not help rural people to better their situation. Joel Cohen describes the trade-off this way: "A metric ton of dung burned as fuel rather than used as fertilizer reduces potential grain output by around 50 kilograms. Around 1985, an estimated 400 million tonnes of dung were burned annually in Africa, Asia, and the Near East. Hence potential food output was reduced by 20 million tons in aggregate, more than 10 kilograms per person in these areas." *How Many People?* 196.

37. In the past, agricultural modernization and urbanization have led to smaller but more productive rural populations. In the United States at the turn of the past century, for example, about a third of the workforce was engaged in agriculture; now less than 3 percent is, yet the country feeds a much larger population and exports large quantities of food. The population of South Korea, one of the original Asian tigers, shifted from 68 percent rural in

1965 to 22 percent rural in 1995 during a period of rapid industrialization and agricultural modernization.

38. Of Earth's remaining species, 18 percent of the mammals, 11 percent of the birds, and 8 percent of the plant species are threatened with extinction as a result of human activities. See Peter Vitousek et al., "Human Domination of Earth's Ecosystems," *Science* 277 (July 1977): 494–499.

Chapter 8   Critical Security Trends

1. James Brook, "Kidnappings Soar in Latin America, Threatening Region's Stability," *New York Times,* 7 April 1995, A8.

2. Philip Howard and Thomas Homer-Dixon, *Environmental Scarcity and Violent Conflict: The Case of Chiapas, Mexico* (Washington, D.C.: American Association for the Advancement of Science; Toronto: University of Toronto, 1996).

3. The mosquito that carries dengue fever bites in the daytime, thus putting at risk many of the outdoor pursuits that attract travelers to the Caribbean.

4. Today, the United States is the sole remaining military superpower. The U.S. military budget is not only larger than those of all U.S. allies put together but more than 7 times as large as that of either Russia or China and more than 100 times as large as that of Iraq. All that spending has produced the world's most powerful military organization. See John Steinbrunner and William Kaufman, "International Security Reconsidered," in *Setting National Priorities,*" ed. Robert Reischauer (Washington, D.C.: Brookings Institution Press, 1997), 155–196. Only the United States has a global reach and the logistical ability to quickly move large numbers of troops and equipment anywhere in the world. Moreover, U.S. military forces have an overwhelming technological advantage. They are far ahead in satellite surveillance, in advanced sensors that allow nighttime operations and all-weather warfare, in more sophisticated communications that give better "command and control," in smart weapons and stealth aircraft, and in antimissile defenses, to name only a few of the tools of modern warfare. These advantages give U.S. units the ability to outmaneuver and overcome larger forces.

In effect, the world now lives under a kind of "Pax Americana." As the Gulf War showed, the United States can overwhelmingly defeat any nation rash enough to oppose it in conventional military conflict. And given present spending patterns, the U.S. advantage will only widen. It would take several decades of systematic military buildup for any conceivable opponent to begin to pose a serious threat. As a result, the risk of conventional warfare has sharply receded—at least wherever the United States and its allies are potentially willing to fight.

5. Jessica Mathews, *Washington Post,* 24 April 1995.

6. Molly Moore and John Ward Anderson, "Key Aide Dismissed in Mexico," *Washington Post*, 3 December 1996, A17.

7. David Hoffman, "Russia's Nuclear Sieve," *Washington Post*, 17 April 1996, A25; A. M. Rosenthal, "Only a Matter of Time," *New York Times*, 22 September 1996, A31.

8. Tom Cochran, an analyst with the Natural Resources Defense Council, assembled these materials in an effort to convince U.S. security officials of the dangers posed by nuclear proliferation. I vividly remember reading through detailed descriptions of such topics as how to machine plutonium and how to shape highly explosive "lenses." The stacks of material he collected, which filled two large filing cabinets in his office, included a crude but workable design for a nuclear bomb designed by a bright high school physics student.

9. Leonard Cole, "The Specter of Biological Weapons," *Scientific American* 275, no. 6 (December 1996): 60–65.

10. General Aleksandr Lebed, former candidate for the presidency of Russia and briefly the top security advisor to President Boris Yeltsin, says bluntly that Russia's efforts to safeguard its nuclear materials are "unsatisfactory." Rosenthal, "Only a Matter of Time."

11. *Washington Post*, 30 November 1996; based on a report from the International Labour Organization.

12. These may include failure to become successful husbands and fathers. *Economist*, 28 September 1996, 24–26.

13. See, for example, Peter G. Peterson, "Facing Up," *Atlantic Monthly* (October 1993).

14. Jean Raspail, *Camp des saints* (The camp of the saints) (Paris: Laffont, 1973), cited in Matthew Connelly and Paul Kennedy, "Must It Be the Rest Against the West?" *Atlantic Monthly* (December 1994).

15. Smugglers operating out of Tangier in Morocco charge $600 per head for those seeking a clandestine boat ride to Spain across the narrow channel at the mouth of the Mediterranean Sea. Marlise Simons, "Tangier a Magnet for Africans Slipping into Spain," *New York Times*, 26 August 1996, A2. Albanian speedboat operators charge similar or even higher prices to ferry illegal immigrants to Italy across the Adriatic Sea. *Economist*, 1 April 1955, 44.

16. Matthew Connelly and Paul Kennedy, "Must It Be the Rest Against the West?"

17. Thomas Homer-Dixon, "Environmental Scarcities and Violent Conflict: Evidence from Cases," *International Security* 19, no. 1 (1994): 5–40.

18. Howard and Homer-Dixon, *Environmental Scarcity and Violent Conflict.*

19. Homer-Dixon, "Environmental Scarcities and Violent Conflict." Cynthia McClintock, "Why Peasants Rebel: The Case of Peru's Sendero Luminoso," *World Politics* 37, no. 1 (1984): 48–84.

20. Miriam Lowi, "West Bank Water Resources and the Resolution of Conflict in the Middle East," Occasional Paper no. 1, Project on Environmental Change and Acute Conflict (September 1992).

21. Ibid.

22. Shengkui Cheng and Xianzhong Ding, "Prospects and Obstacles of Making Development Sustainable in China," in *Proceedings of the Regional Workshop on Sustainable Development in East Asia*, Tokyo, 2–4 October 1996 (Tokyo: Global Industrial and Social Progress Research Institute, 1996).

23. United Nations Population Division, *World Urbanization Prospects: The 1994 Revision* (New York: United Nations, 1995), 87.

24. World Resources Institute, "The Urban Environment," in *World Resources 1996–97* (New York: Oxford University Press, 1996), 1–56.

25. "Latin America's Backlash," *Economist*, 30 November 1996, 15–16 and 19–21.

26. As quoted by Sam Dillon in "Have-Nots Need Stake in Mexico, Envoy Says," *New York Times*, 4 December 1996.

Chapter 9   Critical Social and Political Trends

1. The number of jobs created by women-owned businesses over the past decade in the United States, for example, more than explains the difference between the United States's vigorous expansion of employment and Europe's anemic job creation rate; in effect, the United States has gained a much larger pool of job-creating entrepreneurs as women have moved into this role.

2. World Resources Institute, *World Resources 1994–95* (New York: Oxford University Press, 1994), 52–53.

3. United Nations Development Programme, *Human Development Report 1996* (New York: Oxford University Press, 1996), 69.

4. Lori S. Ashford and Jeanne A. Noble, "Population Policy: Consensus and Challenges," *Consequences* 2, no. 2 (1996): 30.

5. Ibid.

6. United Nations Development Programme, *Human Development Report 1996*, 24.

7. United Nations Development Programme, *Human Development Report* series (New York: Oxford University Press, annual). This U.N. agency gauges progress in human development by such measures as levels of education and literacy, access to health services, and average life span.

8. United Nations Development Programme, *Human Development Report 1996*, 27. The disparities between economic and social development are even more glaring at a country-by-country level. The people of Costa Rica and Argentina, for example, have similar life spans and literacy rates, despite average incomes in Costa Rica less than one-third of those in Argentina. The peo-

ple of Vietnam have longer life spans and higher literacy rates than the people of the Democratic Republic of Congo, despite average incomes less than one-fifth of those in the Congo.

9. Ibid.

10. Judith Bruce, Cynthia B. Lloyd, and Ann Leonard, *Families in Focus* (New York: Population Council, 1995).

11. Ibid., 14–18.

12. Based on field reporting by Casey C. Kelso for the Institute of Current World Affairs, Hanover, New Hampshire, September 1993.

13. See, for example, Helena Norberg-Hodge, *Ancient Futures: Learning from Ladakh* (San Francisco: Sierra Club Books, 1991).

14. Peter Schwartz, *The Art of the Long View* (New York: Doubleday, 1991), 124–140.

15. "The Coming Global Tongue," *Economist*, 21 December 1996, 75–78.

16. Ironically, some U.S. observers have similar concerns about the continuing influx of Latin American immigrants to the Southwest and the rise of Spanish as a second language in that region, fearing dilution of the once dominant Anglo culture.

17. Could such concerns become a kind of conflict—a "clash of civilizations," as a recent and controversial book by Samuel Huntington styles it— between the United States and the "challenger civilizations" of China and Islamic countries? Samuel Huntington, *The Clash of Civilization* (New York: Scribner, 1996). It seems very unlikely. The Islamic countries vary enormously from one another and have never been able to agree on a common agenda that could unite them into a single cultural or political force. As discussed in part IV of this book, anti-Western rhetoric in those countries may really reflect an internal conflict between the forces of tradition (including Islamic fundamentalists) and those of modernization—a conflict whose outcome will shape the Middle East far more than will the region's interaction with the West. China also faces the challenge of modernizing its economy and its technical systems while trying to resist political and cultural change; there, too, resolving that dilemma is primarily an internal struggle, not an external struggle with the West.

18. Jessica T. Mathews, "Power Shift," *Foreign Affairs* 76, no. 1 (1997): 50–66.

19. Ibid.

Chapter 10   Latin America: Equitable Growth or Instability?

1. All economic figures are given in equivalent purchasing power. In 1995, average per capita GNP, used throughout this book as a proxy for average income, was about $5,700 per person.

2. World Bank, *World Development Indicators 1997* (Washington, D.C.: World Bank, 1997).

3. Edwards, *Crisis and Reform in Latin America* (New York: Oxford University Press, 1995), 256–258.

4. Edwards, *Crisis and Reform in Latin America*, 259.

5. Nearly a quarter of the world's emissions of airborne lead occurs in Latin America. See V. Thomas and T. Spiro, "Emissions and Exposure to Metals," in *Industrial Ecology and Global Change*, ed. R. Socolow et al. (New York: Cambridge University Press, 1994).

6. World Resources Institute, *World Resources 1992–93* (New York: Oxford University Press, 1992), 172.

7. Robert Kaiser, "South America's Transformation," *Washington Post*, 15 December 1996, C1–C2.

8. James Brooke, "Home, Home on the Range in Brazil's Heartland," *New York Times*, 26 April 1995, A4.

9. Keith Bradsher, "In South America, Auto Makers See One Big Showroom," *New York Times*, 25 April 1997, C1.

10. "Emerging Market Indicators," *Economist*, 29 March 1997, 116.

11. Edwards, *Crisis and Reform in Latin America*, 309.

12. Peter Gizewski and Thomas Homer-Dixon, *Urban Growth and Violence: Will the Future Resemble the Past?* (Washington, D.C.: American Association for the Advancement of Science, 1995), 5.

13. See chapter 8 for a fuller discussion of the Chiapas uprising.

14. World Resources Institute, *World Resources 1990–91* (New York: Oxford University Press, 1990), 46.

15. Diana Jean Schemo, "Violence Growing in Battle over Brazilian Land," *New York Times*, 21 April 1996, 12; "The Dispossessed," *New York Times Magazine*, 20 April 1997, 42–47.

16. Edwards, *Crisis and Reform in Latin America* (New York: Oxford University Press, 1995), 276.

17. Ibid., 268.

18. Ibid., 312.

19. "Second Wave," *Economist*, 22 March 1997, 96.

20. In Taiwan in the late 1940s, the government imposed ceilings on landholdings and purchased excess land at below-market prices, eliminating the landed elite; one-quarter of the privately held farmland changed hands. South Korea, prodded by the U.S. government during the Korean War, carried out land redistribution that by 1960 had lowered the percentage of farm families who were tenants from 86 percent to 26 percent. The reforms in both of these countries are credited with laying the foundation for more equitable income

distribution and the subsequent development of strong domestic markets. World Resources Institute, *World Resources 1992–93*, 43.

21. Gabriel Escobar, "Bolivia Defies Past, Tries Decentralization," *Washington Post*, 17 September 1995, A25.

22. Alvaro Vargas Llosa, "To Give Latins Real Reform, Start with Property Titles," *Wall Street Journal*, 3 January 1997, A9.

Chapter 11 China and Southeast Asia: Can the Asian Miracle Continue?

1. See, for example, a report by the Australian Department of Foreign Affairs and Trade, *China Embraces the Market: Achievements, Constraints, and Opportunities* (Canberra, Australia: Australian Department of Foreign Affairs and Trade, East Asia Analytical Unit, 1997).

2. Based on field reporting by Cheng Li, a professor of government at Hamilton College, Clinton, New York, for the Institute of Current World Affairs, Hanover, New Hampshire, June 1995.

3. A 1992 survey of nearly 1,500 private entrepreneurs conducted by China's Academy of Social Sciences found that more than 50 percent were former peasants, 47 percent had no more than a middle school education, and nearly 70 percent came from peasant families.

4. These projections assume continuing increases in efficiency. A recent World Bank study concluded that China's energy use will at least triple by the year 2020. World Bank, *China 2020* (Washington, D.C.: World Bank, 1997).

5. Ibid., 75.

6. Edward Cody, "Rise of Market Economy Dents Iron Rice Bowl," *Washington Post*, 1 January 1997.

7. Shengkui Cheng and Xianzhong Ding, "Prospects and Obstacles of Making Development Sustainable in China," in *Proceedings of the Regional Workshop on Sustainable Development in East Asia*, Tokyo, 2–4 October 1996. (Tokyo: Global Industrial and Social Progress Research Institute, 1996).

8. This account is based on a detailed report by Guili Chen, "The Contemporary Era," *Sino-German Journal* 2, no. 55 (20 January 1997) (in Chinese).

9. Lester R. Brown, *Who Will Feed China?* (New York: Norton, 1995). Current total world trade in grain is about 200 million tons per year, about half the amount Brown forecasts will be needed by China alone.

10. These figures include South Korea and Taiwan, which are economically and politically very much a part of Southeast Asia.

11. Charles V. Barber, *The Case Study of Indonesia* (Cambridge, Mass.: American Academy of Arts and Sciences, Committee on International Security Studies, 1997).

Chapter 12   India: A Second Independence?

1. Mahbub ul Haq, *Human Development in South Asia 1997* (New Delhi: Oxford University Press, 1997).

2. Ibid.

3. World Resources Institute, *World Resources 1994–95* (New York: Oxford University Press, 1994), 83–106.

4. World Bank, *Global Economic Prospects and Developing Countries* (Washington, D.C.: World Bank, 1997).

5. World Resources Institute, *World Resources 1994–95*, chapter 5.

6. John F. Burns, "India's Five Decades of Progress and Pain," *New York Times*, 14 August 1997, A1–A7.

7. Madhav Gadgil and Ramachandra Guha, *Ecology and Equity: The Use and Abuse of Nature in Contemporary India* (Routledge, London, 1995).

8. Based on reporting by Kenneth J. Cooper, "Unusual Coalition Aids Poorest in a Key Indian State," *Washington Post*, 30 September 1997. The coalition later fell apart.

9. Gadgil and Guha, *Ecology and Equity*, 170–174.

Chapter 13   Sub-Saharan Africa: Transformation or Tragedy?

1. Ernest J. Wilson, "Globalization, Information Technology, and Conflict in the Second and Third Worlds" (paper presented at a meeting of the Rockefeller Brothers Fund, "Globalization, Inequality, and Conflict in Developing Countries," Pocantico, N.Y., 24–25 April 1997).

2. World Bank, *World Development Indicators 1997* (Washington, D.C.: World Bank, 1997).

3. A recent U.N. report cites more than 5 million refugees from nine such conflicts in recent years in sub-Saharan Africa, not counting the recent civil war in the former Zaire. United Nations Development Programme, *Human Development Report 1966* (New York: Oxford University Press, 1996), 26.

4. M. Hulme et al., *The Effects of Climate Change on Africa* (Stockholm: Stockholm Environment Institute, 1995), 29.

5. The results of the 2050 Project's African workshop were disseminated to participants but not published. The scenarios in this chapter draw on the materials from C. Achebe, G. Hyden, A. Okeyo, and C. Magadza, eds., *Beyond Hunger in Africa: Conventional Wisdom and a Vision of Africa in 2057* (Nairobi, Kenya: Heinemann, 1990).

6. "An African Success Story," *Economist*, 14 June 1997, 47.

7. This scenario is adapted from Achebe et al., *Beyond Hunger in Africa*.

8. Jeffrey Goldberg, "Our Africa," *New York Times Magazine*, 2 March 1997, 33–77.

9. P. Veit, T. Nagpal, and T. Fox, *Africa's Wealth, Woes, Worth* (Washington, D.C.: World Resources Institute, 1995).

10. Ibid., 8.

11. Jeffrey Goldberg, "Our Africa," 76.

Chapter 14   North Africa and the Middle East: Autocracy Forever?

1. J. Mackenzie, *Oil As a Finite Resource: When Is Global Production Likely to Peak?* (Washington, D.C.: World Resources Institute, 1996), 4. As this analysis points out, production from a given oil field declines sharply after about half of its total oil is recovered; the same thing happened with U.S. oil production and is expected to occur again for the world as a whole. Thus, oil is likely to be in increasingly short supply after global production peaks (probably between 2007 and 2019, almost certainly before 2025, if use of oil continues to grow along present trends). From then on, the huge oil fields in Saudi Arabia, Kuwait, Iraq, and Iran will play an increasingly strategic role in world oil supply because there would be no way to replace that production if it were disrupted.

2. *Economist*, 7 June 1997, 42.

3. For the Western reader, it is important to point out the central role religion plays in Islamic culture and in the political institutions of the region. The roles of religion and of the state, rather than being clearly distinct, as in Western culture, overlap in fundamental ways. The Islamic law code called the Shari'ah is a more important guide to family and personal life than are laws passed by the state—and at least in the Sunni Islamic tradition, it allows for no interpretation or adjustment to modern conditions. And no Islamic political leader can afford to ignore the *ulama*, a group of senior religious scholars and jurists. Turkey, a secular Islamic state, is the apparent exception, but here, too, there are strongly fundamentalist factions and a rising religious influence in political matters.

4. Exact figures don't exist because outlaw states such as Libya and Iraq don't report their numbers.

5. These scenarios were originally created by Rob Coppock, a social scientist who was director of the 2050 Project.

6. Unlike most of the region, Iran is part of the Shi'ite Islamic tradition, which allows reinterpretation of Islamic laws by religious leaders and scholars to adapt them to changing conditions, giving the country more social flexibility than in Sunni societies, such as Afghanistan. Thus, should more radical Islamic regimes appear, they are likely to follow Afghanistan's far more rigid approach to Islamic law, hastening the flight of skilled professionals.

7. William H. Lewis, *The Middle East and North Africa: The Military Versus Democracy* (Washington, D.C.: Center for Global Security and Cooperation, 1997).

Chapter 15 Russia and Eastern Europe: Transition to What?

1. Daniel Yergin and Thane Gustafson, *Russia 2010 and What it Means for the World* (New York: Random House, 1993), 201.

2. Cheryl Silver and Dale Rothman, *Toxics and Health* (Washington, D.C.: World Resources Institute, 1995), 35; World Resources Institute, *World Resources 1996–97* (New York: Oxford University Press, 1996), 177–178.

3. "The Endless Winter of Russian Reform," *Economist*, 12 July 1997, 18.

4. In eastern Europe, average incomes range from $9,770 in the Czech Republic to $2,400 in Ukraine.

5. World Bank, *Global Economic Prospects and Developing Countries* (Washington, D.C.: World Bank, 1997).

6. Yergin and Gustafson, *Russia 2010*.

7. Ibid., 187.

8. This scenario is adapted from Yergin and Gustafson, *Russia 2010*. Readers are urged to consult this book for a more detailed version of the scenario and for other relevant scenarios and insights.

Chapter 16 North America, Europe, and Japan: Leadership or Stagnation?

1. World Health Organization, *The World Health Report 1996* (Geneva: World Health Organization, 1996).

2. The population of North America, at 297 million in 1995, is expected to reach 384 million in the year 2050, with a plausible range of 301 million to 452 million. The population of Europe, which here includes the European Union as well as Norway and Switzerland, is expected to decline from 378 million to 341 million (plausible range of 289 million to 377 million). The population of Japan is expected to decline from 125 million to 110 million (plausible range of 96 million to 122 million). The European Union is likely to expand to include a number of countries in eastern and central Europe, increasing the region's population base but reducing its average income; the overall downward trend in population will not change, however, since a similar pattern is occurring in eastern and central Europe.

3. Albert Adriaanse et al., *Resource Flows: The Material Basis of Industrial Economies* (Washington, D.C.: World Resources Institute; Wuppertal, Germany: Wuppertal Institute; The Hague: Netherlands Ministry of Housing, Spatial Planning, and Environment; Tsukuba, Japan: National Institute for Environmental Studies, 1997).

4. In this scenario, governments in virtually every country used markets to ration energy use rather than creating bureaucracies, either by raising energy taxes or by auctioning off energy use permits that could be traded. Likewise, the U.S. legislation setting up energy rationing and those in most other industrial countries required that tax or permit revenues be used to reduce social

security and other employment taxes, recycling the money into the economy—
what economists call a revenue-neutral approach—to minimize the economic
pain of rationing.

5. The ideas in this paragraph are from Peter Schwartz and Peter Leyden,
"The Long Boom," *Wired* (July 1997).

6. Europe may be pursuing its laudable social goals with the wrong policies.
Freer labor markets (so that companies could lay off workers when needed)
and lower employment taxes (making the hiring of new workers less onerous)
would probably create more jobs in the long run—which even at the cost of
some additional short-term unemployment would not threaten Europe's social
goals, however difficult such policies may be politically.

7. There is some economic evidence that greater economic equality facilitates
economic growth; certainly, many of the Southeast Asian economies over the
past several decades have been able to combine both rapid growth and fairly
broad distribution of its benefits.

Appendix

1. United Nations Population Division, *Annual Populations 1950–2050
(The 1996 Revision)*, on diskette (New York: United Nations, 1997); part of
World Population Prospects: The 1996 Revision (New York: United Nations,
forthcoming).

2. World Bank, *World Development Indicators* (Washington, D.C.: World
Bank, 1997).

3. See Paul Raskin and Robert Margolis, *Global Energy in the Twenty-First
Century: Patterns, Projections, and Problems* (Stockholm: Stockholm Environ-
ment Institute, 1995), and Paul Raskin et al., *The Sustainability Transition*
(Stockholm: Stockholm Environment Institute, 1995).

4. Working Group II of the Intergovernmental Panel on Climate Change,
Special Report on the Regional Impact of Climate Change (Geneva: World
Meteorological Organization and United Nations Environment Programme,
1997).

5. *Report of the Secretary-General on a Comprehensive Assessment of the
Freshwater Resources of the World* (New York: United Nations, 1997).

6. Robert Engleman and Pamela LeRoy, *Conserving Land: Population and
Sustainable Food Production* (Washington, D.C.: Population Action International, 1995); Robert Engleman and Pamela LeRoy, *Sustaining Water: Population and the Future of Renewable Water Supplies* (Washington, D.C.: Population Action International, 1993). See also World Resources Institute, *World
Resources 1996–97* (New York: Oxford University Press, 1996), 301–303.

Index